The Correspondence Between
A. A. Markov and A. A. Chuprov
on the Theory of Probability
and Mathematical Statistics

The Correspondence Between
A. A. Markov
and
A. A. Chuprov
on the Theory of Probability
and
Mathematical Statistics

Edited by Kh. O. Ondar

Translated by Charles and Margaret Stein

Springer-Verlag
New York   Heidelberg   Berlin

Professor Charles M. Stein
Department of Statistics
Stanford University
Stanford, California 94305 U.S.A.

Margaret Stein
Stanford, California 94305 U.S.A.

AMS Classification: 01A60 01A99 62-03

Originally published as *O Teorii Veroiatnostei I Matematicheskoi Statistike
(perepiska A.A. Markov u A.A. Chuprov)* by Nauka, Moscow, 1977.

Library of Congress Cataloging in Publication Data

Markov, A. A. (Andreĭ Andreevich), 1856–1922.
  The Correspondence between A. A. Markov and
A. A. Chuprov on the theory of probability and
mathematical statistics.

  Translation of: O teorii veroiatnosteĭ i
matematicheskoĭ statistike/A. A. Markov.
  1. Probabilities. 2. Mathematical
Statistics. 3. Mathematicians—Soviet Union—
Correspondence. I. Chuprov, Aleksandr
Aleksandrovich, 1874–1926. II. Ondar, Kh. O.
III. Title.
QA273.M32913   519.2   81-5767 AACR2

9 8 7 6 5 4 3 2 1

ISBN-13: 978-1-4613-8135-8   e-ISBN-13: 978-1-4613-8133-4
DOI: 10.1007/978-1-4613-8133-4

Introduction

It is a great pleasure to write this Introduction, suggested by the translators and requested by Springer-Verlag. What can be more pleasant than to recall one's student days? My own go back to 1912 when, in Kharkov, I often heard Markov being described as "Neistovyi Andrei." This description is difficult to translate. Perhaps "Andrew the irrepressible," with the addition "who does not pull any punches." Two of the characteristic performances of Markov are briefly described by the editor Ondar. One is Markov's fight "against reaction, backwardness and religion." The other is Markov's renunciation of "all honors and decorations he had received from the tsarist government," this in protest against the exclusion from the Academy of A. M. Gorky, the revered writer. Now I wish to add a third item.

In the tsarist regime, membership in the Imperial Academy was occasionally conferred on high noblemen. Neistovyi Andrei did not like this and, to manifest his disapproval, composed a limerick. It was about a Duke Dundook becoming a member of the Academy, a limerick not suited for the ears of ladies! It is not likely that Markov's limerick was ever published. It circulated by word of mouth.

The little book edited by Ondar is very attractive. Also, I am highly appreciative of the work of the translators Charles and Margaret Stein. They have retained the somewhat antiquated style of writing of Markov and Chuprov: long sentences and special forms of civility. Quite apart from the external form, I find the book very interesting, this not only because of the anecdotal aspect of Neistovyi Andrei pulling no punches in his criticism of Chuprov and others. The historical perspective on the happenings documented in the book, compared with the subsequent developments in probability and mathematical statistics, is most interesting;

certain findings that at the time appeared important did not prove to be significant, and vice versa.

In an effort to estimate the culminating point of the Markov/Chuprov dispute, I contemplate the celebration by the Academy of the bicentenary of the law of large numbers, symbolized by *Ars Conjectandi*, the work of Jacob Bernoulli, published in 1713. In his letter of January 27, 1913 (item 55 of the collection), Markov informs Chuprov that in planning this celebration, he visualizes the possibility of presentations by non-members of the Academy and suggests that he, Chuprov, be one of them. As indicated by the editor Ondar, the response of Chuprov was not found. However, the subsequent letter of Markov (item 56 of the collection) indicates what must have been the contents of the lost letter of Chuprov. In the letter dated January 31 of the same year Markov expresses doubt about the desirability of Chuprov's suggestion to publish a collection of articles by a number of authors, including some from abroad. The reasons for Markov's doubt include the apparent lack of appreciation by the many authors of Bernoulli's work *Ars Conjectandi*. The proposed speakers at the Academy's celebration are Professors A. V. Vasiliev, Chuprov and Markov. In a letter of March 2 (item 58 of the collection) Markov informs Chuprov of the final approval by the Academy of the above plans and invites Chuprov to his home for a joint discussion with Vasiliev.

Actually, the bicentenary celebration was held on December 1, 1913, as planned. Then, in a letter of December 3 (item 62 of the collection), Markov informs Chuprov that, contrary to the suggestion of Vasiliev, he is opposed to the joint publication by the Academy of all three presentations, by Vasiliev, by Chuprov and by himself. Because of the lack of mathematical rigor in Vasiliev's and Chuprov's texts, Markov decided to abstain from publishing his own text! The texts of Markov's and Chuprov's bicentenary presentations are available in Ondar's book as Appendices 3 and 4, respectively.

Now, a few remarks on historical perspectives.

(i) In his many writings, Markov emphasized his mathematical rigor. While the text of his presentation at the Academy's bicentenary celebration is focused on *Ars Conjectandi*, the successive editions of Markov's book *Calculus of Probability* are aimed at the building of the general theory. The Introduction of the third edition, published in 1913 and marked *"Two Hundred Year Jubilee of the Law of Large Numbers,"* is specific in mentioning " ... foundations of the calculus of probability ..." In due course the success of Markov's efforts in this direction became subject to doubt that stimulated other scholars to do better. Without much risk of oversimplification, one might say that modern mathematical theory of probability was born two decades later, due to the work of Academician Andrei Nikolaievich Kolmogorov. Kolmogorov's most inspiring study has the title *Grundbegriffe der Wahrscheinlichkeitsrechnung*. It was published in 1933 by Julius Springer.

Outstanding contributions during the period 1913–1933 may be exemplified as follows: (1) Generalization of the central limit theorem by S. N. Bernstein, my teacher of probability; (2) Harald Cramér's study of 1928; (3) George Pólya's papers on random walk (1921) and on contagion (1930); (4) Émile Borel's Le Hasard (1914); (5) Borel-Cantelli lemma (before 1928); (6) Paul Lévy's characteristic functions and his book of 1925; and (7) Richard von Mises' Theory (1931).

(ii) By the end of his bicentenary speech (Appendix 3 of the Ondar book) Markov refers to the "developments of the law of large numbers that belong already to our time" and mentions "the broadening of the field, in particular, to the extension to DEPENDENT TRIALS and DEPENDENT RANDOM VARIABLES . . ." (caps are mine).

Here, the capitalized terms refer to Markov chains. As is well known, Markov chains generated the theory of Markov processes and, more generally, the theory of stochastic processes that now preoccupies many thousands of scholars all over the world. What a difference from Markov's own opinion about just a broadening of the field of the law of large numbers!

(iii) Preoccupied with the law of large numbers, Markov and (to a degree) Chuprov fail to take notice of a novel subject of study. This subject may be symbolized by the title of a paper by G. T. Fechner *Kollektivmasslehre* published in 1897. The term Kollektivmasslehre may be translated as the study of properties of "collectives," now called "populations." These collectives are supposed to be composed of many entities, all satisfying a certain definition, but differing from each other by some individual characteristics.

The consciousness of populations as an important subject of study is reflected in the famous work of Laplace *Théorie Analytique des Probabilités*. Here, in the edition of 1820, pp. 261–263, Laplace considers two populations of "astronomical entities." One is the population of planets in the solar system. The other is the population of comets. Laplace was interested in the question whether the comets are members of the solar system just like planets, or intruders from the outer space.

When the concept of the Kollektivmasslehre became broadly familiar, there resulted several efforts to develop methodology capable of characterizing the distribution of individual characteristics of population members. The most successful methodology seems to be that of Karl Pearson, commonly known as "Pearson curves." These curves have been sharply criticized by Markov on account of being interpolatory devices, not resulting from a limit theorem on probabilities. This interpolatory character notwithstanding, Pearson curves are useful in many empirical studies.

The attitudes of Markov and Chuprov towards population studies, as reflected in their presentations at the Academy meeting of December 1, 1913, are not identical.

A number of passages in Chuprov's presentation indicate his awareness of populations as important subjects of study. In the very first paragraph

of Chuprov's text (Appendix 4) we read: " . . . interest in collective phenomena." One of subsequent pages contains an even more relevant passage. Here, Chuprov writes about efforts of statisticians to base their studies on complete coverage of a population that is very large. He writes:
"However, even when a complete count is not impossible, statisticians are
beginning more and more to revert to sampling studies, because of the
saving of labor and expense."

Here, I wish to document an achievement of Chuprov not mentioned
in Ondar's book. The following passage is reproduced without change
from the *Journal of the Royal Stat. Soc.,* Vol. CXV (1952), p. 602.

Recognition of priority.—Professor J. Neyman of the University of California writes: I am obliged to Dr. Donovan J. Thompson of the Statistical
Laboratory, Iowa State College, Ames, Iowa, for calling my attention to the
article of A. A. Tschuprow, "On the mathematical expectation of the moments
of frequency distributions in the case of correlated observations" published
in *Metron,* Vol. 2, No. 4 (1923), pp. 646–683, which contains some results
refound by me and published, without reference to Tschuprow, in 1933.

The results in question are the general formula for the variance of the
estimate of a mean in stratified sampling and the formula determining the
optimum stratification of the sample. These formulae appeared first in a Polish booklet *An Outline of the Theory and Practice of Representative
Method, Applied in Social Research* published in 1933 by the Warsaw Institute of Social Problems. Later on they were republished in English in the
Journal of the Royal Statistical Society, Vol. 97 (1934), pp. 558–625. Finally,
the same formulae, again without a reference to Professor Tschuprow, were
given in the second edition of my book, *Lectures and Conferences on Mathematical Statistics and Probability,* Washington, D.C., 1952.

The purpose of this note is, then, to recognize the priority of Professor
Tschuprow, to express my regret for overlooking his results and to thank Dr.
Thompson for calling my attention to the oversight.

P.S. Some readers of the English translation of Ondar's book may find it
convenient to be reminded that the following four names refer to the same
city: St. Petersburg, Petersburg, Petrograd and Leningrad. The city's original name, St. Petersburg, was changed to the Russian name, Petrograd,
in 1914. After the death of Lenin in 1924 the city was renamed Leningrad.

Acknowledgment. The present *Introduction* was prepared using the
facilities of the Statistical Laboratory, University of California, Berkeley,
with partial support of the Office of Naval Research [N000 14 75 C 0159]
and of the Army Research Office [DA AG 29 79 0093]. All the opinions
expressed are those of the author.

JERZY NEYMAN
Statistical Laboratory
University of California, Berkeley

Translators' Remarks

After Professor Neyman had reviewed the Russian edition of this correspondence, he suggested to us and to Springer-Verlag that this historical record might also be of interest to some English-speaking readers. The work of translation, although difficult, was rewarding. We are grateful to Professor Neyman for suggesting this translation and for kindly consenting to write an introduction to it.

Those interested in the history of probability and statistics will appreciate the work of the editor, Kh. O. Ondar, in the time-consuming and difficult task of searching out and deciphering the handwritten correspondence of Markov and Chuprov. Ondar's preface helps us to understand the historical setting for this correspondence.

With only two changes of any importance we have tried to translate the text as it stands. Part of the footnote to Letter No. 10 has been rewritten, using modern terminology and notation. In addition the editor's review of the correspondence has been substantially shortened as indicated at that point. His valuable introduction has been retained. The other changes are merely corrections of a few misprints and minor errors that would detract from the readability of the text.

We are indebted to Jerzy Neyman and to Persi Diaconis for reading the translation and making some useful suggestions. Of course any remaining errors and clumsiness are the responsibility of the translators alone.

January 1981

CHARLES M. STEIN
MARGARET D. STEIN
Stanford, California

Preface

Andrei Andreevich Markov and Alexander Alexandrovich Chuprov are outstanding Russian scientists whose merits are recognized throughout the world.

Andrei Andreevich Markov was born on July 14, 1856 in the province of Ryazan in the family of a middle civil servant. In the 1860's the family moved to Petersburg where Markov received his secondary education. His studies at the gymnasium, based on the study of dead, ancient languages, were not very fascinating to the young student. He received mediocre grades and complaints were often sent from the gymnasium about his poor progress. However, an exception was mathematics in which Markov took a great interest very early, having already begun the independent study of higher mathematics in the gymnasium.

In 1874 Markov entered the University of Petersburg. His student years coincided with the highest point of P. L. Chebyshev's scientific and pedagogical activity. Chebyshev, who was the founder of the Petersburg mathematical school and an outstanding Russian scientist, had a decisive influence on the development and the direction of the young student's scientific interests. In 1878 Markov received a gold medal for his scientific work "On the integration of differential equations with the aid of continued fractions." In that same year he was graduated from the university and was retained there to prepare for a career as professor.

In 1880 Markov defended his master's thesis, and four years later his doctor's dissertation. Markov's scientific merits were so high that the Academy of Sciences elected him an Adjunct Member in 1886, an Extraordinary Member four years later, and an Academician after another six years.

All of Markov's further life was wholly devoted to science. He pre-

sented his last scientific work to the Academy of Sciences only several
months before his death on July 20, 1922.

The name of Markov is connected with major discoveries in the fields
of number theory, the approximation of functions, the problem of
moments, the calculus of finite differences, etc. However, without doubt
his numerous works on the theory of probability occupy a central place in
his extensive legacy. Developing the ideas of his great teacher Chebyshev,
Markov obtained a series of fundamental results which determined the
further direction and development of the science and promoted the transformation of the theory of probability into one of the most powerful tools
of research of modern science. Already the theme of academician Markov's first papers was closely connected with the fundamental laws of the
theory of probability—the law of large numbers and the central limit
theorem.

Chebyshev gave a general outline of the proof of the central limit theorem, using the method of moments which he originated. However, quite
a few difficulties arose in Chebyshev's outline of a proof and these were
not easy to overcome. Markov devoted a long series of works to the solution of these problems.

In Markov's work the classical formulation of the theory of probability
achieved its well-known perfection and further development. A particularly important contribution to science was Markov's development of the
theory of chains of random processes, now called "Markov chains." The
theory of chains represents a mathematical model of a process without
after effects which describes a physical system when the probability of
transition to another state depends only on the state of the system at a
given time and not on the preceding history of the process. Markov exten-
sively developed the theory of this scheme of stochastic processes, extending the basic laws of the classical schemes to it. His work became the starting point for the creation of the general theory of stochastic processes, one
of the most fruitful branches of contemporary probability theory.

It is also important to mention the work of Markov on some problems
of mathematical statistics.

Of great interest is the justification of the method of least squares,
bringing full clarity to the estimation of the error of the results of adjust-
ment of given observations. Markov introduced and subsequently made
systematic use of new, important ideas equivalent to the modern ideas of
unbiased and efficient statistics for estimation of constant parameters of
distribution laws on the basis of samples.

All of Markov's mathematical creativity was close in style and spirit to
that of Chebyshev: the problems were formulated in wide generality and
the solutions were carried out to the point of obtaining numbers and
algorithms, with careful attention to detail and to clarity and accuracy of
exposition. The scientific ideas of Markov gave mathematicians a program
of research for many years to come.

While engaged in scientific work and teaching, Markov also vigorously commented on social issues and was an active fighter against reaction, backwardness and religion. In 1902 a wave of protest arose in the country when A. M. Gorky was excluded from the ranks of honorary academicians. The progressive-minded people of Russia came to the defense of the great writer. Markov could not stand aside from this movement. As a sign of protest he renounced all honors and decorations he had received from the tsarist government.

Alexander Alexandrovich Chuprov was born on February 18, 1874 in the city of Mosalsk[1] but grew up and was educated in Moscow where his father A. I. Chuprov was a professor at Moscow University. After receiving his primary education at home and finishing the gymnasium in 1892, Chuprov enrolled at the physico-mathematical faculty at Moscow University. In 1896 he was graduated successfully, having written his thesis on "The theory of probability as the foundation of theoretical statistics."

In order to study social sciences Chuprov went first to the University of Berlin and later to the University of Strasbourg, and after completing his studies there he immediately began to prepare for a master's examination at the faculty of law at Moscow University. In the spring of 1902 he passed his examinations and became a teacher in the economics department at the Petersburg Polytechnic Institute.

In May, 1909 Chuprov's *Essays on the Theory of Statistics* was published. It was presented at Moscow University as a master's dissertation which he defended on December 2. However, the doctor's degree was immediately conferred upon its author.

Chuprov's book *Essays on the Theory of Statistics* opened up the enormous significance of this science to Russian statisticians, it showed the depth of its philosophical and logical foundations, widened scientific horizons, and gave an orderly, well-developed introduction to the theory of statistics. In 1914 the professional journal *Statisticheskiĭ Vestnik* was founded. After the completion of his Essays on the Theory of Statistics, a definite change in the scientific interests of Chuprov occurred, and although he also kept his interest in the logical aspects of statistical methods in his further research, mathematical problems came all the more to the forefront. In 1911–1917 Chuprov worked intensely on mathematical questions. His scientific correspondence with academician Markov took place during this time.

During the whole period of teaching at Petersburg Polytechnic Institute, Chuprov went abroad regularly at vacation time to work in the great scientific libraries. In June, 1917 he, as usual, went to Sweden at vacation time to study material at the Principal Statistical Bureau in Stockholm, which he used in his extensive work in connection with the First World War. He had already begun this work in the previous year.

It is clear from his letters that he intended to return to Petrograd toward the beginning of the academic year, i.e. around September, 1917, but at first his illness prevented it, and later—also monetary difficulties.

In September, 1918 he wrote from Stockholm that he planned to arrive in Petrograd in two weeks in order to continue his work at the institute, Not receiving any money from Petrograd, he found himself in a difficult financial situation in Stockholm, and wrote about this in his letters.

From this time no more letters arrived from Chuprov and it is not known why he did not return to Petrograd. We may suppose that it was because of the lack of necessary funds since the passage by steamship from Stockholm to Petrograd cost 3000 rubles, according to the information in the newspapers.

"Professor A. A. Chuprov is one of the most eminent statisticians, not only in Russia but in all of Europe, where his work on theoretical questions in statistics is well known to all specialists. One of the special aims of the Polytechnic Institute is to train statisticians. Hence, it is extremely important for them to have in their midst such an outstanding representative of Russian statistical science as Professor Chuprov, whose name and academic work are known to every Russian statistician. From A. A. Chuprov's letters it is clear that while living in Stockholm he indeed worked persistently on the scientific work he had undertaken on the abovementioned questions. It is essentially ready for the press and he will publish it immediately upon his return to Russia.

The services of Professor A. A. Chuprov to statistical science determined the advancement of his candidacy to the post of head of the Central Statistical Department of the Soviet Republic which was established at this time. In April, 1918 Comrade Elizarov, the commissar for insurance matters, made the official offer to him."[2]

Experiencing financial difficulties, Chuprov, in January, 1919, took the position of director of the statistical bureau of the prerevolutionary Central Union in Stockholm and was in charge of the publication *Bulletin of World Economy*. But after a year and a half he decided to devote himself completely to the scientific research that had been interrupted. Chuprov left his work at the Central Union and in the middle of 1920 moved to Dresden where he lived in strict seclusion, absorbed in scientific research. Chuprov disapproved of the activity of the emigrant circles who were hostile in attitude to the Soviet state and avoided them completely. He delayed his return to his homeland until the completion of the series of scientific works he had begun. V. I. Bortkiewicz, in an obituary of Chuprov, wrote: "It was not because of any personal fears that he did not want to return to his native land; on the contrary, he had good reason to expect consideration and even affectionate regard for himself."

In the period 1918–1925 Chuprov published an enormous number of papers of about 70 printed pages. In 1925 his health deteriorated. In September, 1925 he went to Rome to a meeting of the International Statistical Institute with a paper on sample surveys. In Rome Chuprov became seriously ill and died on April 19, 1926 at the age of 52.

Being a follower of the school of the Russian mathematicians P. L. Chebyshev and A. A. Markov, Chuprov considered it necessary that the construction of the stochastic theory of statistics be based on a foundation of strict mathematical concepts and methods. At this time the ideas of V. Lexis were thought to contradict those of British empiricism. However, unlike V. Bortkiewicz who considered the positions of Lexis and Pearson irreconcilably incompatible, Chuprov brought forward the idea of their synthesis, that is, the basing of Pearson's constructions on the mathematically rigorous methods developed by Chebyshev and Bienaymé. Chuprov saw the primary objective of his scientific work as the logical and mathematical foundation of statistical methodology.

Chuprov showed exhaustively how the fluctuations of statistical data depend, firstly, on the degree of constancy and variability of the probabilities on which they are based, and secondly, on the presence of dependence between the separate observations and on its character. On the basis of the theory of stability he determined necessary and sufficient conditions for the applicability of the law of large numbers. Applying the method of mathematical expectation to the working out of the theory of stability, and carefully distinguishing between the a priori (theoretically given) characteristics of an aggregate and the a posteriori (based on the observations), Chuprov came to the conclusion that Lexis's criterion for stability is not always applicable and he pointed out the restrictions underlying the results based on this criterion.

In recognition of the great merits of Chuprov in the development of science, he was elected a Corresponding Member of the Russian Academy of Sciences, a correspondent of the Royal Economic Society in London, a member of the International Statistical Institute, and later (in 1923) an honorary member of the Royal Statistical Society in London.

From 1910 to 1917 Markov and Chuprov carried on a lively correspondence on a number of questions of the theory of probability and mathematical statistics. More than 100 of the letters of these scholars have not been published heretofore. These letters contain an enormous amount of material from which one can judge not only the way the problems of the mathematical expectation of Lexis's criterion (Q) and the mathematical expectation of moments were worked out, but also the evolution in the thinking of both scholars, their polemic fervor, sources of the statements of a series of problems, and their mutual influence on scientific methodology. This correspondence opens up to the contemporary reader the historical development of the theory of probability and mathematical statistics in Russia.

We discovered the letters of A. A. Markov in the Department of Rare Books and Manuscripts of the Gorky Scientific Library at Moscow State University.[3] The letters of A. A. Chuprov are kept in the Leningrad section of the Archives of the Academy of Sciences of the U.S.S.R.[4] They were found with the assistance of A. A. Markov, corresponding member of the Academy of Sciences of the U.S.S.R. and son of academician A. A. Markov. It is regrettable that not all of the letters of Chuprov have been preserved. Part of them, apparently, are lost (for example, for the year 1912 only one letter was found, and not one for 1913 and 1915, although the correspondence between the scholars continued in this period).

A. A. Markov and A. A. Chuprov lived and worked in Petersburg, corresponding by mail. The dates of the letters are preserved and refer to the old calendar.

Academician B. V. Gnedenko and Professor K. A. Rybnikov were very helpful in composing the commentary on the letters of Markov and Chuprov, in the review of their correspondence, and also in the preparation of this book for the press.

KH. O. ONDAR

---

[1] Karpenko, B. I., The Life and Work of A. A. Chuprov, *Scientific Writings on Statistics,* V. 3, Moscow, Press of the Academy of Sciences of the U.S.S.R., 1957.

[2] State archives of the Great October socialist revolution and the socialist construction of the Leningrad region, reserve 3121, inventory 12, storage unit 7–81, sheets 164–165.

[3] The A. A. Chuprov reserve, carton 23, storage unit 1.

[4] Reserve 173, inventory 1, storage unit 23.

Contents

Introduction .... *v*

Translators' Remarks .... *ix*

Preface .... *xi*

The Correspondence Between A. A. Markov and
A. A. Chuprov 1

A Review of the Correspondence Between A. A. Markov
and A. A. Chuprov
KH. O. ONDAR 136

Appendices: A. A. Markov and A. A. Chuprov on the Law of
Large Numbers 147

    Appendix 1. On the Basic Principles of the Calculus of
    Probability and on the Law of Large Numbers
    A. A. MARKOV 149

    Appendix 2. A Review of A. A. Markov's Book, *The Calculus*
    of Probability
    A. A. CHUPROV 154

    Appendix 3. The Bicentennial of the Law of Large Numbers
    A. A. MARKOV 158

    Appendix 4. The Law of Large Numbers in Contemporary
    Science
    A. A. CHUPROV 164

The Correspondence Between
A. A. Markov and
A. A. Chuprov

No. 1

(Postcard from Markov to Chuprov)

2 November 1910

I note with astonishment that in the book of A. A. Chuprov, *Essays on the Theory of Statistics,* on page 195, P. A. Nekrasov, whose work in recent years represents an abuse of mathematics[1], is mentioned next to Chebyshev.

A. MARKOV

---

[1] Markov has in mind the following place in Chuprov's book, *Essays on the Theory of Statistics* (published in 1909, p. 195): "... the mathematical constructions, beginning with the original theorem of Bernoulli in its mathematically simple form in *Ars Conjectandi* as well as in the elegant formulation by Laplace, and culminating in the more general "law of large numbers" of Poisson, the still more general theorem of Chebyshev, and the most general version of all, constructed by Professor Nekrasov and Bruns, are separated by a gulf from the law of large numbers which establishes a connection between probabilities of events and their relative frequencies."

Markov is right that Nekrasov is referred to undeservedly here. In 1898 Nekrasov published the article "General properties of independent events in connection with the approximate computation of functions of very large numbers" (*Matematicheskiĭ Sbornik*, 1898, v. XX). Apparently Chuprov's opinion is explained by Nekrasov's unfounded claim that his findings allegedly extend the results of Chebyshev and Markov. In this connection Markov wrote the short but very sharp comment. The fact is that while claiming to extend Chebyshev's theorem, Nekrasov proved nothing in the article. At the end of the article he writes: "A detailed derivation of all the foregoing results will be presented by me later if circumstances permit me to put my computations into a form suitable for publication." (*Matematicheskiĭ Sbornik*, 1898, v. XX, p. 442).

Indeed the formulas introduced by Nekrasov without proofs were absolutely unfounded, as Markov showed persuasively. (A. A. Markov, "Reply to a note of P. A. Nekrasov," *Izvestiia Fiziko-Matematicheskovo Obshchestva pri Kazanskom Universitete,* 2nd series, 1899, v. 9, No. 3).

As is clear from his reply Chuprov agreed with Markov's opinion and noted that he (Chuprov) was speaking "only of the degree of generality in the formulation of the problem" (Editor's note).

No. 2

(Letter from Chuprov to Markov)

4 November 1910

Highly esteemed Andrei Andreevich:

I shall not argue with your evaluation of P. A. Nekrasov's scientific work in recent years; in this my opinions do not differ much from yours. However, I cannot recognize that it would be fair to remain silent on the matter in which his name is mentioned in my book. It seems to me that it is impossible to deny that Nekrasov had some role in developing the formulation of the question of the relation between the probabilities of events and their frequencies. But in mentioning him next to Chebyshev, do I really put him on the same level of talent with Chebyshev? I am speaking only of the degree of generality in the formulation of the problem. The order of names on this point does not in any measure coincide with an appraisal in terms of mathematical interest or statistical importance. As far as importance for statistical theory is concerned, I, for example, place not only Nekrasov but also Bruns and even Chebyshev himself below Poisson. But on the other hand, from the point of view of mathematics, I would rank Poisson after Chebyshev. These are three quite different scales.

The second edition of my *Essays on the Theory of Statistics* will be out in a few days. Page 195 is there without change. However, in view of the sympathetic recommendation which Nekrasov gave me in his note in connection with Bortkiewicz's review of my *Essays*, I considered it my responsibility to touch also, in the new edition, on the "abuse" of mathematicians in Nekrasov's work. However, not finding it appropriate to expound at length for the few to whom Nekrasov's reflections are known and for the fewer to whom they are interesting, I confined myself to a critique of those inferences having to do with the meaning of stability of functions of statistical data to which Nekrasov's lame attempt at a logical analysis of the hypotheses of Chebyshev's theorem lead.

Respectfully yours,

A. Chuprov

No. 3

(Letter from Markov to Chuprov)

6 November 1910

My dear Alexander Alexandrovich:

Of course I was also surprised by your reference to Bruns whom I consider a negligible quantity.

I can judge all work only from a strictly mathematical point of view and from this viewpoint it is clear to me that neither Bruns nor Nekrasov nor Pearson[1] has done anything worthy of note. You speak about some kinds of most general constructions, but I cannot find these constructions in their work.

Meanwhile I do find highly general theorems from authors whom you have entirely forgotten: A. M. Liapunov[2] and A. A. Markov. The unique service of P. A. Nekrasov, in my opinion, is namely this: he brings out sharply his delusion, shared, I believe, by many, that independence is a necessary condition for the law of large numbers. This circumstance prompted me to explain, in a series of articles, that the law of large numbers and Laplace's formula can apply also to dependent variables[3]. In this way a construction of a highly general character was actually arrived at, which P. A. Nekrasov can not even dream about.

I considered variables connected in a simple chain and from this came the idea of the possibility of extending the limit theorems of the calculus of probability also to a complex chain.

Independence is not required for the application of these theorems, but on the other hand it is necessary to assume existence of certain constant quantities. This existence is already assumed by the theory and therefore it is impossible to deduce this from the theory. And so I will stick to my opinion that your reference to Bruns and Nekrasov is wrong, as long as you do not cite for me their general constructions.

With complete respect,

A. MARKOV

---

[1] It is impossible to agree with such a characterization of the scientific merits of Bruns and especially of Pearson. Pearson's outstanding work in the field of theoretical statistics is universally recognized today (Editor's note).

[2] A. M. Liapunov (1857–1918)—an eminent Russian mathematician and specialist in theoretical mechanics, academician, prominent representative of the Petersburg mathematical school established by Chebyshev.

[3]The discussion is about the well-known work of A. A. Markov in which he carries out the mathematical investigation of a special class of dependent events, called "chains" by the author. Later these dependencies became known in science as "Markov chains." It is clear from the text of the letter that Nekrasov's errors led Markov to think about the problem of dependent events (Editor's note).

No. 4

(Letter from Chuprov to Markov)

10 November 1910

Highly esteemed Andrei Andreevich:

Returning today from a brief holiday I received your letters and the reprint. I am very grateful to you for them. I wrote my *Essays* abroad and foreign literature received an advantage over the Russian in them: I did not examine our journals (particularly the provincial ones) systematically and your article and Liapunov's remained unknown to me. With lively interest I read through the article[1] you sent me and keenly regret that I had not received either it or your other references a month earlier when, by delaying publication, I could have repaired this important gap in the second edition of my book. I have just added a presentation of your point of view—namely, that independence is not a necessary condition for the law of large numbers—to the second edition, only by references to the work of Bohlmann and Bachelier[2].

As for the scheme which I set forth on page 276, I also do not see any difference in principle between it and the case of drawings from one urn (cf. page 340 of my book)[3]. I am speaking of it only as a generalization of the way in which the scheme of a single urn is ordinarily reflected in presentations by statisticians. That point of view, in which the two schemes are represented as one and the same, is grappled with by statisticians with very great difficulty; a new striking example of this is the recent article by Fr. Edgeworth (a paper on the application of the theory of probability to statistics, presented at the Paris meeting of the International Statistical Institute) who again entangles himself in the interpretation of this scheme.

Sincerely yours,

A. CHUPROV

[1]Here, apparently, the discussion concerns Markov's article "Investigation of the general case of trials connected in a chain," *Zapiski A. N. po Fiz-Mat. Otd.*, Series VIII, 1910, v. 25, No. 3 (Editor's note).

[2]Chuprov, A. A., *Essays on the Theory of Statistics*, 3rd edition, M., Gosstatizdat, 1959, pp. 231–232 (Editor's note).

[3]Chuprov is speaking about the following urn scheme: there are s urns; let p_i be the probability of drawing a white ball from the ith urn ($i = 1, 2, \ldots, s$). It is assumed that before each drawing the urn from which the ball is to be drawn is determined by chance, while the probability that the ball will be drawn from the first urn remains the same at all times and is equal to p_1'; the probability of drawing the ball from the second urn is always equal to p_2'; and so forth. The probability that the ball will be drawn from the sth urn is equal to p_s'. If, under such conditions, returning the ball that was drawn to that urn from which it was drawn, we carry out n successive drawings and repeat this experiment μ times, then in the row of μ relative frequencies of white corresponding to each of the series, everything will be just as if the balls were drawn from the one urn with such a number of white and black balls in it that the probability of drawing from it a white ball would equal $p = p_1'p_1 + p_2'p_2 + p_3'p_3 + \cdots + p_s'p_s$. (Editor's note).

No. 5

(Letter from Markov to Chuprov)

11 November 1910

My dear Alexander Alexandrovich:

I most humbly beg you to point out to me those articles of Bohlmann and Bachelier to which you refer. Up to now I had thought that I was the first to dwell on the application of the law of large numbers (as a theorem of the calculus of probability) to dependent variables and I noted the important cases where such an application appeared possible[1]. Incidentially, let me remark that the second edition of my book *The Calculus of Probability* came out in 1908. I learned only accidentally about your book from Professor Dehn[2].

With complete respect,

A. MARKOV

[1]Markov, A. A., "Extension of the law of large numbers to dependent variables," *Izvestiia Fiziko-Matematicheskovo Obshchestva pri Kazanskom Universitete*, 2nd series, 1906, v. 15, No. 4 (Editor's note).

[2]E. V. Dehn (1867–1933)—Professor of economic geography, Petersburg Polytechnic Institute (Editor's note).

No. 6

(Letter from Chuprov to Markov)

12 November 1910

Highly esteemed Andrei Andreevich:

The work of Bohlmann, which to me seems just as interesting for the result obtained as for the method of obtaining it, is entitled "Die Grundbegriffe der Wahrscheinlichkeitsrechnung in ihrer Anwendung auf die Lebensversicherung" and appeared in *Atti del IV Congresse Internazionale dei matematici,* v. III, Roma, 1910. If you do not have the proceedings of the congress at hand, I can make my reprint available to you.

Bachelier's article "Théorie des probabilités continues" was published in *Journal des mathematiques,* v. LXXI (series VI, v. II), Paris, 1906. The generalized point of view toward the law of large numbers to which you adhere and on which I base my work on the theory of statistics causes difficulties even for mathematicians. In this connection the new edition of Czuber's *Wahrscheinlichkeitsrechnung und ihre Anwendung* is interesting. In the second volume Czuber referred several times to one of my German articles[1], and although he quotes several portions of it, not without sympathy, he apparently cannot make up his mind to emphasize the basic point of view (cf., for example, pp. 46–47, footnote).

Sincerely yours,

A. CHUPROV

---

[1]F. Czuber's work was published in 1899 in Leipzig. The discussion is about Chuprov's article "Die Aufgaben der Theorie der Statistik," (*Jahrbuch für Gesetzgebung, Verwaltung und Volkswirtschaft,* 1905, Bd. XXIX, H.2). Russian translation: Chuprov, A. A., "Questions of statistics," M. Gosstatizdat, 1960, pp. 43–90 (Editor's note).

No. 7

(Letter from Markov to Chuprov)

15 November 1910

My dear Alexander Alexandrovich:

Thank you for the reference that you sent me. I of course have seen Bachelier's article but I strongly dislike it. I do not attempt to judge its significance for statistics but with respect to mathematics, it has no importance in my opinion. In any case, it does not contain an extension of Bernoulli's theorem to dependent variables.

Also I do not find such a generalization in Bohlmann's article, which I did not know of. Bohlmann led me to recall something that I had noted formerly in lectures and that P. A. Nekrasov has noted, namely, that in order to consider the mathematical expectation of a squared sum, pairwise independence of the variables is sufficient. Corresponding to this, the dispersion, in the sense of Lexis, which does not go beyond the mean squared deviation, can be shown to be normal[1] for dependent trials if they are pairwise independent. Believing that this can be of interest to you I shall give an example. Let all the odd trials $1, 3, 5, \ldots, 2i - 1, 2i + 1,$ $\ldots$ be independent of each other and let the probability of event E for each of them be $1/2$.

For the even trials we shall assume that event E must occur on the $2i$th trial if the results of the $(2i - 1)$st and $(2i + 1)$st trials are the same, but it does not occur on the $2i$th trial if the results of the $(2i - 1)$st and $(2i + 1)$st trials are different. Our trials are pairwise independent. And therefore the dispersion in the sense of Lexis must be normal although the complete independence of the trials is out of the question.

In this case, letting x_l be the number 1 or 0 corresponding to the occurrence or non-occurrence of event E at the lth trial and putting $x_l - 1/2 = y_l$, we have independent variables

$$y_1, y_3, y_5, \ldots, \ldots$$

and the variables associated with them

$$y_{2i} = x_{2i} - \tfrac{1}{2} = 2(x_{2i-1} - \tfrac{1}{2})(x_{2i+1} - \tfrac{1}{2}) = 2y_{2i-1}y_{2i+1}.$$

Then the mathematical expectation of each y_l and the product $y_l \cdot y_k$ for $l \neq k$ is equal to zero, and the mathematical expectation of $y_l^2 = 1/4 = 1/2 \cdot 1/2$.

Therefore the mathematical expectation of $(x_1 + x_2 + \cdots + x_n - n/2)^2 = E(y_1 + y_2 + \cdots y_n)^2 = n/4 = n \cdot 1/2 \cdot 1/2$.

The difference between this case and the normal is found only by considering the mathematical expectation of higher powers.

With complete respect,

A. MARKOV

---

[1]Here it seems appropriate to keep the word 'normal' which is used in the Russian text. Of course it does not refer to the normal (Gaussian) distribution (Translator's note).

No. 8

(Letter from Chuprov to Markov)

15 November 1910

Highly esteemed Andrei Andreevich:

Your example is extraordinarily interesting for statistics. It is instructive in two respects. First of all, of course, because you constructed it: it brings out boldly that independence is not a necessary condition even for normal dispersion (in the sense of Lexis), not to speak of the law of large numbers in general, and that consequently normality of dispersion is not firm evidence of independence. This is all the more important because it is practically impossible to bring such matters to the attention of statisticians without such an intuitive example, as I have had the opportunity to convince myself. And, on the other hand, it is very easy to see from your example how normal dispersion is attained here, despite the lack of independence, because of the mutual compensation of the moments increasing dispersion against the norms, and those decreasing it. In this connection it seems to me to be interesting, for greater clarity, to compare your example with the following two variants.

1. As in your example, the odd trials are independent of each other and the probability of event E for each of them is 1/2. However, for the even trials the situation is as follows: if for both adjacent odd trials the event E occurs, then it also occurs for the included even trial; if on both adjacent odd trials it does not occur, then it also fails to occur on the even trial; if the event E occurs on one of the adjacent odd trials but

not on the other, then the probability of occurrence on the even trial as well as the probability of non-occurrence is 1/2.

2. The odd trials, as before, are independent. However, for the even trials the situation is the following: if event E occurs on both adjacent odd trials, then it does not occur on the included even trial; if for the (adjacent) odd trials it does not occur, then it does occur on the even one; finally, if it occurs on one of the adjacent odd trials but not on the other, then on the even trial its probability of occurrence as well as non-occurrence is 1/2.

The first version gives supernormal dispersion, the second version, supernormal stability; the additional term in Bohlmann's formula ($\Sigma(p_{ik} - p_i p_k)$) is equal to $(5n - 1)/8$ in the first case (if there are $2n + 1$ trials); in the second (case) it is equal to $-(3n + 1)/8$. In your basic example the stability is normal. Thus, it is clear what is happening here and I think even a blind man could see it.

Thank you very much for the reprints of your articles.

Sincerely yours,

A. CHUPROV

No. 9

(Letter from Markov to Chuprov)

17 November 1910

Highly respected Alexander Alexandrovich:

I am very pleased that my example excited your curiosity. I must confess that in constructing it I made a mistake. I imagined that in the given case one could explain the irregularity of the dispersion by considering higher order moments, but this is not justified. The principal terms of the moments of all orders turn out to be exactly the same for this case as for the case of independent trials. Therefore, the asymptotic expression of the probability remains in the form of Laplace's integral, as if our trials were completely independent of each other. Of course the probabilities of the different assertions about the number of occurrences of the event in this case cannot coincide with the probabilities of such assertions in the normal case.

As long as we consider a small number of trials (I went up to only 20 trials here), we can observe several peculiarities of this case.

However, all of these peculiarities are smoothed out when, as usual we consider, for a very large number of trials, the probability that the ratio of the number of occurrences of the event to the number of trials lies within given bounds. As for your modified examples, the mathematical expectation of the square, according to your calculations, increases in proportion to the number of trials, and therefore Bernoulli's theorem remains valid. One must assume that the asymptotic expression for the probability also holds.

In connection with the asymptotic expression for the probability, I consider it necessary to observe that up to now no one anywhere has succeeded in coming up with any precise formula other than Laplace's formula. On the contrary, Laplace's formula was established by Liapunov and me under such general assumptions for independent variables as well as dependent that it should apply to all the usual cases. And so I cannot attach any importance to the concoctions of Pearson and others establishing some kind of asymmetric expressions[1].

One can consider all of these formulas only as purely interpolative, devoid of a firm theoretical basis.

Apropos of this, let me remind you again that in all of these arguments the existence of certain constants[2] plays a central role. If statistics throws light on the existence of such constants, then this fact is quite remarkable, and the reference to the calculus of probability does not alter the matter because we do not know why the probability should remain constant.

I note that the more complicated a scheme you devise the more difficult it is to recognize it as legitimate. As for the "free will" which P. A. Nekrasov likes to talk about, the reference to it cannot be useful. I do not wish to enter into an argument about free will (one can use this word out of ignorance of the causes of events), but I should mention that it can only be a factor working against legitimacy: the constancy of some number or other occurs, not by virtue of freedom but because of some constant cause.

With complete respect,

A. Markov

---

[1] We have already mentioned that it is impossible to agree with Markov's opinion of Pearson's scientific merit (Editor's note).

[2] Among all of the families of curves which have been proposed in order to make it possible for statisticians to treat mathematically various sorts of data, those that have been most justified in practice seem to be Pearson's curves. Their justification ordinarily lies in the fact that in some approximate sense the normal distribution, the binomial distribution, and the hypergeometric distribution all satisfy a differential equation of the form:

$$\frac{1}{\Omega} \cdot \frac{d\Omega}{dx} = \frac{a + x}{b + cx + ex^2}$$

for certain values of the constants a, b, c and e or parameters entering into the equation of the curve of the distribution. It is undoubtedly something like this that Markov has in mind (Editor's note).

No. 10

(Letter from Chuprov to Markov)

17 November 1910

Highly esteemed Andrei Andreevich:

Pearson's formulas, which P. A. Nekrasov mentions in his note in connection with Bortkiewicz's review of my book, certainly have no relation whatever to the questions we are studying, and Nekrasov's rebuke that I did not pursue Laplace's formulas further seems to me, to put it simply, devoid of meaning. It makes me happy that, as I consider myself entitled to conclude from your letter, you also concur.

One place in your preceding letter is not entirely clear to me. You write that you do not find that Bohlmann has extended Bernoulli's theorem to dependent variables. On the contrary, the formula in Bohlmann's[1] theorem XIII seems to me an important generalization. In it, for example, are included those cases which you consider in the article "Extension of the law of large numbers to dependent variables." Today I did several of the problems which you looked into, and for problems §2 and 4, proceeding from Bohlmann's formula, arrived at the same solutions that you indicated[2].

Sincerely yours,

A. CHUPROV

---

[1] The discussion here concerns the following formula of Bohlmann (XIII) for squared error of the arithmetic mean:

$$M^2 = \sum_{i=1}^{n} p_i q_i + \sum_{i \neq k} (p_{ik} - p_i p_k),$$

which was referred to above (Editor's note).

[2] In §2 of Markov's article the following problem is investigated. Let us consider variables $x_1, x_2, \ldots, x_n, \ldots$ with corresponding mathematical expectations $a_1, a_2, \ldots, a_n,$ $\ldots$ and for brevity let us put $x_i - a_i = z_i$.

First consider the case where $x_1 + x_2 + \cdots + x_n$ is the number of occurrences of an event A in n successive trials connected in such a way that the conditional probability of A given the past is p' if A occurred on the preceding trial and p'' if A did not occur on the preceding trial, the unconditional probability of A being p. Note that the numbers p, p', and p'' are related among themselves by the simple equality $p = pp' + (1 - p)p''$, from which, if we are given two of these numbers it is not difficult to find the third:

$$p = \frac{p''}{1 - p' + p''}, \quad p' = 1 + p'' - \frac{p''}{p}, \quad p'' = p\frac{1 - p'}{1 - p}.$$

Markov proved that

$$E(z_1 + z_2 + \cdots + z_n)^2 < \frac{np(1 - p)(1 + p' - p'')}{1 - p' + p''}$$

if $p' > p''$. It will then follow that the law of large numbers is applicable to this case. Markov's result can be obtained easily from Bohlmann's formula.

The computations can be carried out in the following way. First we verify by induction that, for $j > 0$,

$$p_{i,i+j} = P(A_i \cdot A_{i+j})$$
$$= p[p + (p' - p)(p' - p'')^{j-1}].$$

This is true for $j = 1$ since

$$p_{i,i+1} = pp' = p[p + (p' - p)].$$

Assuming it true for a particular j, say $j = j_o$, and writing $A^c_{i+j_o}$ for the event that A does not occur on the $(i + j_o)$th trial we obtain

$$p_{i,i+j_o+1} = P(A_i \cdot A_{i+j_o+1})$$
$$= P(A_i \cdot A_{i+j_o} \cdot A_{i+j_o+1}) + P(A_i \cdot A^c_{i+j_o} \cdot A_{i+j_o+1})$$
$$= p\{[p + (p' - p)(p' - p'')^{j_o-1}]p' + [(1 - p) - (p' - p)(p' - p'')^{j_o-1}]p''\}$$
$$= p(p' - p)(p'' - p')^{j_o} + p[pp' + (1 - p)p''] = p[(p' - p)(p'' - p')^{j_o} + p].$$

This completes the proof by induction.

Then

$$\sum_{i \neq k} (p_{ik} - p_i p_k) = 2\sum_{i=1}^{n-1} \sum_{k=i+1}^{n} (p_{ik} - p^2) = 2\sum_{i=1}^{n-1} \sum_{j=1}^{n-i} p(p' - p)(p' - p'')^{j-1}$$
$$= 2\sum_{i=1}^{n-1} p(p' - p)\frac{1 - (p' - p'')^{n-i}}{1 - p' + p''}$$
$$= \frac{2p(p' - p)}{1 - p' + p''}\left[(n - 1) - \frac{1 - (p' - p'')^{n-1}}{1 - p' + p''}(p' - p'')\right]$$
$$= 2p(p' - p)\left[\frac{n}{1 - p' + p''} - \frac{1 - (p' - p'')^n}{(1 - p' + p'')^2}\right].$$

It follows from Bohlmann's formula that

$$M^2 = np(1 - p) + 2p(p' - p)\left[\frac{n}{1 - p' + p''} - \frac{1 - (p' - p'')^n}{(1 - p' + p'')^2}\right]$$
$$< \frac{npq(1 + p' - p'')}{1 - p' + p''}, \text{ if } p' > p''.$$

In §4, Markov looks at an example to which the law of large numbers is inapplicable. Let $x_1, + \cdots + x_n$ be the number of white balls among n balls drawn from an urn under the following conditions:

1. The initial number of white balls in the container is a, and the number of other balls is b.
2. Each ball drawn from the container goes back into the container together with another ball of the same color. With the help of the method of mathematical expectation Markov proved that

$$E(z_1 + z_2 + \cdots + z_n)^2 = \frac{nab(n + a + b)}{(a + b)^2 (a + b + 1)}.$$

It is easy to obtain this result from Bohlmann's formula.
In fact, in the given problem it is not difficult to convince ourselves that

$$p_1 = p_2 = \cdots = p_n = \frac{a}{a + b};$$

$$p_{ik} = P(A_i \cdot A_k) = \frac{a(a + 1)}{(a + b)(a + b + 1)}, \text{ where } k \neq i.$$

$$p_{ik} - p_i p_k = \frac{ab}{(a + b)^2(a + b + 1)};$$

$$\sum_{i \neq k} (p_{ik} - p_i p_k) = n(n - 1) \frac{a\,b}{(a + b)^2(a + b + 1)};$$

$$\sum_{i=1}^{n} p_i q_i = n \frac{a}{a + b} \cdot \frac{b}{a + b} = n \frac{a\,b}{(a + b)^2}.$$

Consequently, from Bohlmann's formula

$$M^2 = n \frac{ab}{(a + b)^2} + \frac{abn(n - 1)}{(a + b)^2(a + b + 1)} = \frac{nab(n + a + b)}{(a + b)^2(a + b + 1)}.$$

(Editor's note, abbreviated by the translators.)

Translators' note: This is a special case of the Pólya urn scheme. The general case is treated more thoroughly with statistical applications in Eggenberger and Pólya, "Über die Statistik verketteter Vorgänge, *Zeitschrift für angewandte Mathematik und Mechanik*, v. 3, 1923, pp. 279–289.

No. 11

(Letter from Markov to Chuprov)

18 November 1910

Highly respected Alexander Alexandrovich:

On page 350[1] you mention the apparently fashionable formula of Bortkiewicz-Lexis which plays an important role in their theories. You did not derive this formula but you indicated the conditions under which such a formula holds. You refer to Czuber[2] but in one of your letters to me you yourself spoke not very approvingly about him.

In my opinion this formula has not been proved and cannot be proved, and is sustained only by respect for authority.

First of all, the formation of the groups is very strange. According to your explanation the trials belonging to one group are independent of each other. But for dependent trials all of the theoretical calculations are significantly more complicated than for independent, which such calculators as Czuber apparently did not observe.

If the trials are divided into groups only according to the value of the probability, then nothing prevents us from breaking them up, not into μ but into $n\mu$ groups; then p is reduced to one and the expression M' turns out to be equal to M.

In order not to make unsubstantiated statements I shall give a simple example. Assume two groups of trials, each composed of two trials. For each group one decides by lot whether to draw a ball from the container with one white or, on the contrary, with one black ball. The die gives a probability of 1/2 for each case. Here four cases arise which are given in Table 1. The table shows that the mathematical expectation of the sum of squares of deviations from the mean is equal to:

$$\tfrac{1}{4}\{\tfrac{1}{4} + \tfrac{1}{4} + \tfrac{1}{4} + \tfrac{1}{4} + \tfrac{1}{4} + \tfrac{1}{4} + \tfrac{1}{4} + \tfrac{1}{4}\} = \tfrac{2}{4}$$

If, however, we consider 4 independent trials for which the probability of drawing a white ball is 1/2, then we have 16 cases which can be broken down into the categories indicated in Table 2. This table gives, for the mathematical expectation of the sum of squares of deviations from the mean:

$$\tfrac{1}{2}\left(\tfrac{1}{16} + \tfrac{1}{16} + \tfrac{1}{16} + \tfrac{9}{16}\right) + \tfrac{3}{8}\left(\tfrac{1}{4} + \tfrac{1}{4} + \tfrac{1}{4} + \tfrac{1}{4}\right) = \tfrac{3}{4}.$$

Table 1

Probability	Number of white balls drawn				Mean	Deviation from the mean			
$\tfrac{1}{4}$	1	1	1	1	1	0	0	0	0
$\tfrac{1}{4}$	1	1	0	0	$\tfrac{1}{2}$	$+\tfrac{1}{2}$	$+\tfrac{1}{2}$	$-\tfrac{1}{2}$	$-\tfrac{1}{2}$
$\tfrac{1}{4}$	0	0	1	1	$\tfrac{1}{2}$	$-\tfrac{1}{2}$	$-\tfrac{1}{2}$	$+\tfrac{1}{2}$	$+\tfrac{1}{2}$
$\tfrac{1}{4}$	0	0	0	0	0	0	0	0	0

Table 2

Balls	Mean number	Probability	Deviation from the mean			
4 white	1	$\frac{1}{16}$	0	0	0	0
3 white, 1 black	$\frac{3}{4}$	$\frac{1}{4}$	$\frac{1}{4}$	$\frac{1}{4}$	$\frac{1}{4}$	$\frac{3}{4}$
2 white, 2 black	$\frac{1}{2}$	$\frac{3}{8}$	$\frac{1}{2}$	$\frac{1}{2}$	$\frac{1}{2}$	$\frac{1}{2}$
1 white, 3 black	$\frac{1}{4}$	$\frac{1}{4}$	$\frac{3}{4}$	$\frac{1}{4}$	$\frac{1}{4}$	$\frac{1}{4}$
no white	0	$\frac{1}{16}$	0	0	0	0

In the second assumption, as you see, a larger number is obtained than in the first. The simple example which I have produced proves the unfoundedness of the concoctions of Lexis and Bortkiewicz[3].

Yours, with complete respect,

A. MARKOV

---

[1]Chuprov, A. A., *Essays on the Theory of Statistics,* M., Gosstatizdat, 1959, pp. 281–282.

[2]Czuber, Emanual (1851–1925)—German mathematician.

[3]As Chuprov remarks in his answering letter, the example pointed out by Markov "does not contradict the constructions of Lexis and Bortkiewicz." In fact, the example worked out by Markov is not correct, i.e., it is not what Lexis and Bortkiewicz had in mind since they took deviations not from the mean of each line of the table (each line consists of four possible outcomes of the trials) but from the mathematical expectation with respect to all possible combinations of them, i.e., from 1/2. If this is worked out in agreement with the constructions of Lexis-Bortkiewicz, then we obtain the following result:

Variant I

Probability of series	No. of white balls drawn in series	Deviation from general mean	Squared deviation	Squared deviation × Probability of series
$\frac{1}{4}$	4	+2	4	$4 \cdot \frac{1}{4} = 1$
$\frac{1}{4}$	2	0	0	0
$\frac{1}{4}$	2	0	0	0
$\frac{1}{4}$	0	−2	4	$4 \cdot \frac{1}{4} = 1$

| | Mean $= \frac{8}{4} = 2$ | | | Mathematical expectation of squared deviation $E\sigma_1^2 = 2$ |

Variant II

Probability of series	No. of white balls drawn in series	Deviation from general mean	Squared deviation	Squared deviation × Probability of series
$\frac{1}{16}$	4	$+2$	4	$4 \cdot \frac{1}{16} = \frac{1}{4}$
$\frac{1}{4}$	3	$+1$	1	$1 \cdot \frac{1}{4} = \frac{1}{4}$
$\frac{3}{8}$	2	0	0	0
$\frac{1}{4}$	1	-1	1	$1 \cdot \frac{1}{4} = \frac{1}{4}$
$\frac{1}{16}$	0	-2	4	$4 \cdot \frac{1}{16} = \frac{1}{4}$
	Mean $= \frac{16}{8} = 2$			Mathematical expectation of squared deviation $E\,\sigma_2^2 = 1$

In this way $\sigma_2^2 < \sigma_1^2$, just as it should be according to Lexis-Bortkiewicz (Editor's note).

No. 12

(Postcard from Markov to Chuprov)

18 November 1910

In my opinion in the last letter you did not express yourself quite precisely. The cases I indicated are NOT included in Bohlmann's cases, but contain them as *special* cases. There is a huge difference. I am prepared to admit that Bohlmann gave an elegant special formula, but he did not point out even one *new* (after my article) case of the generalization of Bernoulli's theorem.

Yours,

MARKOV

No. 13

(Postcard from Markov to Chuprov)

18 November 1910

Dwelling on the question concerning the sum of squares of the deviation from the mean number of occurrences of the event μ, I come to the conclusion that your scheme cannot correspond to Lexis's formula. If we assume that the trials are independent of each other and the probabilities differ, then a formula resembling Lexis's formula is obtained

$$E \sum \left(x_i - \frac{x_1 + x_2 + \cdots + x_n}{n} \right)^2$$
$$= (n-1)\, p\, (1-p) + \frac{\sum (p_i - p)^2}{n},$$

where $p_i = E\, x_i = p_i \cdot 1 + (1 - p_i) \cdot 0$ and

$$p = \frac{p_1 + p_2 + \cdots + p_n}{n}.$$

Yours,

A. MARKOV

No. 14

(Letter from Chuprov to Markov)

19 November 1910

Highly esteemed Andrei Andreevich:

I also do not especially like the formula I introduced on page 350[1], not because it is false but rather because it is awkward: the variables p_I, p_{II}, $\ldots$, p_μ entering into it are not given directly by the conditions of the problem. I did not consider it convenient to by-pass this formula since I set myself the task of facilitating, as far as possible, for the statistician-reader the further study of the question in whatever statistical literature he may have in which he may come across this formula. However, I do not

consider this formula basic, but rather the one on page 357. The deriva-
tion of this last formula was given by Bortkiewicz in *The Law of Small
Numbers* in the second appendix. Now I prefer to derive it starting from
Bohlmann's formula.

The example which you have indicated does not contradict either this
formula or the constructions of Lexis and Bortkiewicz in general. You did
not compute that sum of squares which the statisticians attracted by that
formula have in view. You assume that two groups of two trials are combined for the determination of the relative frequency. Your schematic
table considers that the density in each such series of four trials can, with
one and the same probability of 1/4, take values: 1, 1/2, 1/2, 0. Under your
conditions the mathematical expectation of the relative frequency is 1/2
but the expectation of the sum of squares of deviations which interests
statisticians is $1/4(1/4 + 0 + 0 + 1/4) = 1/8$. We obtain the same value
1/8 by using the formula on page 357.

In the case of independence the sum of squares that we are discussing
is equal to 1/16, i.e., it is not larger but smaller than the first.

My attitude toward Czuber is this: I acknowledge that his textbook is
very useful but I consider the author an untalented man and fairly muddle-headed; he makes blunders that are almost incomprehensible.

Sincerely yours,

A. Chuprov

---

[1]Chuprov assumes here as earlier that we have a series of urns with different propor-
tions of white and black balls. Here the probability of selecting the ith urn for the trials
is g_i and the probability of removing a white ball from it is c_i. "If, under the conditions
described, we repeat the experiment μ times, then, as earlier, we shall obtain a sequence
of μ relative frequencies of white. The mean frequency for all the μ in the series is, as the
mathematical theory shows, not far from $c_0 = g_1 c_1 + g_2 c_2 + \cdots + g_s c_s$ as its most probable value. The variations of the different frequencies from the mean will exceed the
normal level and moreover, will exceed it more strongly the larger k is, where k denotes
the number of different cases connected to each other in the composition of the series
being considered; namely, the modulus for the series of μ relative frequencies under consideration turns out to be equal to

$$\sqrt{\frac{2c_0(1 - c_0)}{n} + \frac{2(k - 1)}{n} \cdot \alpha^2}$$

where α^2 denotes the quantity $g_1(c_1 - c_0)^2 + g_2(c_2 - c_0)^2 + \cdots + g_s(c_s - c_0)^2$." (See
Chuprov, A. A., *Essays on the Theory of Statistics*, M., Gosstatizdat, 1959, pp. 281–282.)
(Editor's note).

No. 15

(Letter from Markov to Chuprov)

19 November 1910

My dear Alexander Alexandrovich:

Yesterday I indicated my conviction that the Bortkiewicz-Lexis formula has not been proved by anyone and can hardly be proved. In any case, as is apparent from a simple example, it does not conform to your scheme.

At the same time I observed, in a postcard, what is in my opinion an important shortcoming in all statisticians; they are satisfied with half an explanation.

Before proceeding to a theoretical calculation it is necessary to define clearly the thing being calculated and to establish its correspondence with that which is calculated in practice. I am convinced that statisticians care little about this. Suitable standards, in my belief, are included at the end of my book (the method of least squares).

To my regret however, I have often heard that my presentation is not sufficiently clear. Allow me to call your attention to the calculations concerning the determination of the probability of an event from observations.

Assume that μ series of n trials are carried out. Let the number of occurrences of an event in series (i) be equal to

$$x_i^{(1)} + x_i^{(2)} + \cdots + x_i^{(n)},$$

so that the estimated probability corresponding to this series is

$$\frac{x_i^{(1)} + x_i^{(2)} + \cdots + x_i^{(n)}}{n}.$$

I shall call it ξ_i. We have a sequence of numbers $\xi_1, \xi_2, \ldots, \xi_\mu$. Considering the deviations of these numbers from their arithmetic mean

$$\xi = \frac{\xi_1 + \xi_2 + \xi_3 + \cdots + \xi_\mu}{\mu},$$

let us compute the sum of squares

$$W = (\xi_1 - \xi)^2 + (\xi_2 - \xi)^2 + \cdots + (\xi_\mu - \xi)^2.$$

This is the quantity that one usually computes. In practice we can deal only with particular values of this variable, and in theory—only with its mathematical expectation. If in place of ξ we put the mathematical expectation of ξ which I shall denote by the letter p, and consider

$$(\xi_1 - p)^2 + (\xi_2 - p)^2 + \cdots + (\xi_\mu - p)^2,$$

then this will already be incorrect because we cannot obtain such a variable from our numbers. Of course in theory we can consider the mathematical expectation of this quantity, but such a consideration will not correspond to that with which we are concerned in practice. And so we shall concern ourselves with the definition of the mathematical expectation of W, where we shall start from as general a point of view as possible.

Let $p_i^{(j)}$ denote the probability that $x_i^{(j)}$ will be 1, and $q_i^{(j)} = 1 - p_i^{(j)}$ the probability that it will be 0. If you wish you can assume in the final inference that

$$p_i^{(1)} = p_i^{(2)} = \cdots = p_i^{(\mu)}.$$

However, it is difficult to imagine that such an assumption is correct. If a variation in the variables $p_i^{(j)}$ exists, then one must admit that it is unlikely that we have succeeded in combining the observations with identical probabilities into groups.

We have

$$W = \sum_i (\xi_i - \xi)^2 = \sum \xi_i(\xi_i - \xi) - \xi \sum (\xi_i - \xi)$$

$$= \sum_i \xi_i^2 - \mu\xi^2 = \sum_i \xi_i^2 - \frac{\sum \xi_i^2 + 2\sum \xi_i \cdot \xi_k}{\mu}.$$

Therefore

$$W = \frac{\mu - 1}{\mu} \sum \xi_i^2 - \frac{2}{\mu} \sum \xi_i \xi_k,$$

and since

$$\xi_i^2 = (\xi_i - p_i + p_i)^2 = (\xi_i - p_i)^2 + 2p_i(\xi_i - p_i) + p_i^2,$$

then

$$E\, \xi_i^2 = E\left\{ \frac{\xi_i' + \xi_i'' + \cdots + \xi_i^{(n)}}{n} \right\}^2 = E\, \frac{\left[\sum_j \{\xi_i^{(j)} - p_i^{(j)}\} \right]^2}{n^2} + p_i^2$$

$$= p_i^2 + \frac{\sum_j p_i^{(j)} q_i^{(j)}}{n^2} = p_i^2 + \frac{\sum_j \{p_i^{(j)} - (p_i^{(j)})^2\}}{n^2}$$

and

$$E\,W = \frac{\mu-1}{\mu}\Sigma p_i^2 + \frac{\mu-1}{\mu}\cdot\frac{\sum\limits_{i,j} p_i^{(j)}\,q_i^{(j)}}{n^2} - \frac{2}{\mu}\sum\limits_{i,k} p_i p_k.$$

Let us consider a transformation of this expression. We begin with the sum:

$$\omega_i = p_i' - (p_i')^2 + p_i'' - (p_i'')^2 + \cdots + p_i^{(n)} - (p_i^{(n)})^2$$
$$= \sum_j p_i^{(j)} - \sum_j (p_i^{(j)})^2.$$

We have

$$\sum_j (p_i^{(j)} - p_i)^2 = \sum_j (p_i^{(j)})^2 - n p_i^2,$$

and therefore

$$\sum_j (p_i^{(j)})^2 = n p_i^2 + \sum_j (p_i^{(j)} - p_i)^2 \qquad \text{and}$$
$$\omega_i = n p_i - n p_i^2 - \sum_j (p_i^{(j)} - p_i)^2.$$

Adding the ω_i for $i = 1, 2, 3, \ldots, \mu$, we obtain

$$\sum_{i,j} p_i^{(j)} \cdot q_i^{(j)} = n \cdot \sum (p_i - p_i^2) - \sum_{i,j} (p_i^{(j)} - p_i)^2.$$

Further, we take into account some simple identities:

$$2 \sum_{i,k} p_i p_k = \left(\sum p_i \right)^2 - \sum p_i^2 = \mu^2 p^2 - \sum p_i^2$$

and

$$\sum p_i^2 = \sum (p_i - p + p)^2 = \sum (p_i - p)^2 + \mu p^2.$$

In this way we obtain

$$E\,W = \frac{\mu-1}{\mu}\cdot\frac{\sum (p_i - p_i^2)}{n} - \frac{\mu-1}{\mu}\cdot\frac{\sum\limits_{i,j} (p_i^{(j)} - p_i)^2}{n^2}$$
$$+ \frac{\mu-1}{\mu}\sum p_i^2 + \frac{1}{\mu}\sum p_i^2 - \mu p^2$$

$$= \frac{\mu - 1}{\mu} \cdot \frac{\sum (p_i - p_i^2)}{n} - \frac{\mu - 1}{\mu} \cdot \frac{\sum\limits_{i,j} (p_i^{(j)} - p_i)^2}{n} + \sum p_i^2 - \mu p^2$$

$$= \frac{\mu - 1}{\mu} \cdot \frac{\sum (p_i - p_i^2)}{n} - \frac{\mu - 1}{\mu} \cdot \frac{\sum\limits_{i,j} (p_i^{(j)} - p_i)^2}{n^2} + \sum (p_i - p)^2.$$

If we have $p_i^{(j)} = p_i$, then we obtain

$$E\,W = \frac{\mu - 1}{\mu} \cdot \frac{\sum (p_i - p_i^2)}{n} + \sum (p_i - p)^2.$$

I have not yet carried the calculation to its conclusion; it is still necessary to reorganize the sum $\Sigma(p_i - p_i^2)$.

For this one must again make use of the equality

$$\Sigma p_i^2 = \mu p^2 + \Sigma(p_i - p)^2.$$

From this we have

$$\Sigma(p_i - p_i^2) = \mu(p - p^2) - \Sigma(p_i - p)^2.$$

Substituting this expression in the general formula we obtain

$$E\,W = (\mu - 1) \cdot \frac{p - p^2}{n} - \frac{\mu - 1}{\mu} \cdot \frac{\sum\limits_{i,j} (p_i^{(j)} - p_i)^2}{n^2}$$
$$+ \frac{1}{\mu} \sum (p_i - p)^2.$$

For $p_i^{(j)} = p_i$ our final equality

$$E\,W = (\mu - 1) \cdot \frac{p - p^2}{n} + \frac{1}{\mu} \sum (p_i - p)^2$$

resembles the formula of Bortkiewicz-Lexis somewhat but does not coincide with it. In the general case we still have a series of terms of opposite sign (for $\mu = 1$ my formula gives $W = 0$, as it should). The assumption that $p_i^{(j)} = p_i$ is extremely implausible and therefore it is impossible to neglect the elements of opposite sign.

I hope that there is not any kind of computational error in my calculations. I hope also that I have presented the question clearly and properly. Czuber's computations seem to me to be devoid of a firm basis.

Yours, with complete respect,

A. MARKOV

No. 16

(Letter from Markov to Chuprov)

19 November 1910

Highly respected Alexander Alexandrovich:

It is evident that I left out one term and therefore I will repeat the computation which differs from Czuber's computation only in the quantity being computed and in its generality:

$$\xi_i = \frac{x'_i + x''_i + \cdots + x_i^{(n)}}{n}, \quad \xi = \frac{\xi_1 + \xi_2 + \cdots + \xi_\mu}{\mu},$$

$$W = \Sigma(\xi_i - \xi)^2, \qquad p_i = \frac{p'_i + p''_i + \cdots + p_i^{(n)}}{n},$$

$$p = \frac{p_1 + p_2 + \cdots + p_\mu}{\mu}.$$

According to Czuber $W = \Sigma(\xi_i - p)^2$. But it is impossible to compute this.

$$W = \Sigma(\xi_i - \xi)^2 = \Sigma\xi_i(\xi_i - \xi) - \xi\,\Sigma(\xi_i - \xi)$$

$$= \Sigma\xi_i^2 - \xi\,\Sigma\xi_i = \Sigma\xi_i^2 - \mu\xi^2.$$

$$\xi_i^2 = (\xi_i - p_i)^2 + 2p_i(\xi_i - p_i) + p_i^2.$$

$$E\,\xi_i^2 = E\,(\xi_i - p_i)^2 + p_i^2.$$

$$E\,\xi^2 = E\,(\xi - p)^2 + 2p \cdot E\,(\xi - p) + p^2 = p^2 + E\,(\xi - p)^2$$

$$= p^2 + \frac{1}{\mu^2}\,E(\xi_1 - p_1 + \xi_2 - p_2 + \cdots + \xi_\mu - p_\mu)^2$$

$$= p^2 + \frac{1}{\mu^2}\sum_i E(\xi_i - p_i)^2.$$

$$E\,W = \sum p_i^2 - \mu p^2 + \frac{\mu - 1}{\mu}\sum_i E\,(\xi_i - p_i)^2.$$

$$\xi_i - p_i = \frac{1}{n}\sum_j (x_i^{(j)} - p_i^{(j)}).$$

$$E\,(\xi_i - p_i)^2 = \frac{1}{n^2}\sum_j E\,(x_i^{(j)} - p_i^{(j)})^2 = \frac{\displaystyle\sum_j p_i^{(j)} - \sum_j (p_i^{(j)})^2}{n^2}.$$

$$\{p_i^{(j)}\}^2 = (p_i^{(j)} - p_i)^2 + 2p_i(p_i^{(j)} - p_i) + p_i^2.$$

$$\sum_j \{p_i^{(j)}\}^2 = \sum (p_i^{(j)} - p_i)^2 + np_i^2.$$

$$\Sigma p_i^2 = \Sigma(p_i - p)^2 + \mu p^2.$$

$$E\,(\xi_i - p_i)^2 + \frac{\sum\limits_j p_i^{(j)} - \sum\limits_j (p_i^{(j)})^2}{n^2} = \frac{p_i - p_i^2}{n} - \frac{\sum\limits_j (p_i^{(j)} - p_i)^2}{n^2}.$$

$$E\,W = \sum (p_i - p)^2 + \frac{\mu - 1}{\mu} \sum \frac{p_i - p_i^2}{n} - \frac{\mu - 1}{\mu n^2} \sum_{i,j} (p_i^{(j)} - p_i)^2$$

$$= \sum (p_i - p)^2 + \frac{\mu - 1}{n}(p - p^2) - \frac{\mu - 1}{\mu n} \sum (p_i - p)^2$$

$$- \frac{\mu - 1}{\mu n^2} \sum_{i,j} (p_i^{(j)} - p_i)^2$$

$$= \frac{\mu - 1}{n}(p - p^2) + \left(1 - \frac{\mu - 1}{\mu n}\right) \sum (p_i - p)^2$$

$$- \frac{\mu - 1}{\mu n^2} \sum_{i,j} (p_i^{(j)} - p_i)^2.$$

This formula reveals the possibility of subnormal dispersion (in the sense of Lexis and Bortkiewicz) for independent trials.

Yours,

A. MARKOV

The case where all the p_i are the same but the numbers $p_i^{(j)}$ are different is interesting. Then the formula shows a dispersion lower than normal.

No. 17

(Postcard from Markov to Chuprov)

19 November 1910

I computed the mathematical expectation of $\Sigma(\xi_i - \xi)^2$ because this quantity should be computed; however, the general nature of the result does not change if we take $\Sigma(\xi_i - p)^2$.

$$\Sigma(\xi_i - \xi)^2 = \Sigma(\xi_i - p + p - \xi)^2$$
$$= \Sigma(\xi_i - p)^2 + 2(p - \xi)\,\Sigma(\xi_i - p) + \mu(p - \xi)$$
$$= \Sigma(\xi_i - p) - \mu(\xi - p)^2.$$

$$E \, \Sigma(\xi_i - p)^2 = E \, \Sigma(\xi_i - \xi)^2 + \mu \, E \, (\xi - p)^2$$

$$= \sum p_i^2 - \mu p^2 + \frac{\mu - 1}{\mu} \sum E \, (\xi_i - p_i)^2$$

$$+ \frac{1}{\mu} \sum E \, (\xi_i - p_i)^2$$

$$= \Sigma p_i^2 - \mu p^2 + \Sigma E \, (\xi_i - p_i)^2$$

$$= \sum_i (p_i - p)^2 + \sum_i \frac{\sum_j p_i^{(j)} - \sum_j (\mathrm{p}_i^{(j)})^2}{n^2}$$

$$= \sum (p_i - p)^2 - \frac{1}{n^2} \sum_{i,j} (p_i^{(j)} - p_i)^2 + \frac{1}{n} \{\Sigma p_i - \Sigma p_i^2\}$$

$$= \frac{\mu}{n} (p - p^2) + \left(1 - \frac{1}{n} \right) \sum (p_i - p)^2 - \frac{1}{n^2} \sum_{i,j} (p_i^{(j)} - p_i)^2.$$

Yours,

A. MARKOV

No. 18

(Postcard from Markov to Chuprov)

19 November 1910

I find Czuber's computations correct but not corresponding to reality. It is necessary to consider not the variable

$$- \frac{m_i}{s} + p$$

where p is unknown, but the variable

$$- \frac{m_i}{s} + p_0 = - \frac{m_i}{s} + \frac{m_1 + m_2 + \cdots + m_r}{r \cdot s}.$$

For that reason, of course, it is impossible not to allow variations in the components of the separate numbers p_1, p_2, ..., p_r, if the question of a change in the probability arises.

Yours,

A. MARKOV

No. 19

(Letter from Markov to Chuprov)

20 November 1910

My dear Alexander Alexandrovich:

Various things are grouped together in your last letter. Firstly, you refer me to page 357[1] about which there was no discussion. Regarding page 357, after the computations on the separate pages which you quoted, I am ready to acknowledge the correctness of your results except for just one point. On this point you refer to other statisticians, but I consider such a method of proof unconvincing. In order to examine deviations from the mathematical expectation it is necessary to know it.

Indeed, it always remains unknown and is replaced by the arithmetic mean of the numbers obtained. Therefore, statisticians compute not squared deviations from the mathematical expectation but squared deviations from the mean number which they obtained. I also compute the mathematical expectation of such a variable with which statisticians really have to deal. Here of course one may replace the arithmetic mean of the numbers obtained by the mathematical expectation in those cases where one quantity differs little from the other.

In the example I cited, a statistician may obtain four combinations:

$$1, \quad 1, \quad 1, \quad 1,$$
$$1, \quad 1, \quad 0, \quad 0,$$
$$0, \quad 0, \quad 1, \quad 1,$$
$$0, \quad 0, \quad 0, \quad 0.$$

In the first combination he finds the probability equal to one and the corresponding sum of squares equal to zero. In the last combination the probability turns out to be zero and the sum of squares of the deviations is also zero. And only in the second and third combinations does the statistician obtain the value 1/2 and the sum of squares of deviations become 1. Since all four combinations are equally likely the mean value of the sum turns out to be 1/2. One cannot obtain the numbers you indicated for the first and last combinations because the value of the probability found in these cases will not be 1/2 but 1 and 0.

We shall not speak of the formula on page 350 if you yourself do not attach great importance to it; I only observe that you refer to Czuber, but his probabililites are assigned without chance. Independent of that, which I consider a mistake of statisticians, I must turn your attention to another circumstance, hoping that in this case you agree with me. The fact is that although your schemes are very interesting, they give rise to doubt. In what way can such groups with constant probability be chosen? Let us permit variation in the different groups. Of course your schemes, to all appearances, are not adaptable to this. But the simpler schemes of Czuber permit this very easily. Then you see that for independent trials the dispersion can easily turn out to be lower than normal, which according to Bortkiewicz and Lexis is impossible. And this is possible even in the case where instead of the true sum of squares one takes an impossible one as statisticians do. I want to call your attention to this circumstance.

With complete respect,

A. MARKOV

---

[1]Chuprov, A. A., *Essays on the Theory of Statistics*, M., Gosstatizdat, 1959, p. 239.

No. 20

(Supplement to the Letter of 20 November 1910 from Markov to Chuprov)

For simplicity everything is multiplied by n^2.

$$W = \Sigma(x_i' + x_i'' + \cdots + x_i^{(n)} - np)^2$$

$$= \Sigma\{x_i' + x_i'' + \cdots + x_i^{(n)} - p_i' - p_i'' - \cdots - p_i^{(n)}$$
$$+ p_i' + p_i'' + \cdots + p_i^{(n)} - np\}^2$$

$$= \Sigma(x_i' + x_i'' + \cdots + x_i^{(n)} - p_i' - p_i'' - \cdots - p_i^{(n)})^2$$
$$+ 2(p_i' + p_i'' + \cdots + p_i^{(n)} - np) \Sigma(x_i' + x_i'' + \cdots + x_i^{(n)}$$
$$- p_i' - p_i'' - \cdots - p_i^{(n)}) + (p_i' + p_i'' + \cdots + p_i^{(n)} - np)^2;$$

$$E\,W = \sum_{i,j} \{p_i^{(j)} - (p_i^{(j)})^2\} + \sum_i (p_i' + p_i'' + \cdots + p_i^{(n)} - np)^2.$$

$$\sum_j (p_i^{(j)})^2 = \sum_j (p_i^{(j)} - p_i)^2 + np_i^2;$$

$$E\,W = \sum_i np_i - \sum_i np_i^2 - \sum_{i,j} (p_i^{(j)} - p_i)^2 + \sum n^2(p_i - p)^2.$$

$$\sum_i p_i^2 = \sum (p_i - p)^2 + \mu p^2;$$

$$E\ W = n^2 \sum_i (p_i - p)^2 - n \sum_i (p_i - p)^2 - \sum_{i,j} (p_i^{(j)} - p_i)^2 + \mu n(p - p^2)$$
$$= \mu n(p - p^2) + (n^2 - n) \sum (p_i - p)^2 - \sum_{i,j} (p_i^{(j)} - p_i)^2.$$

Here I have computed the expression one usually considers in theory, which does not correspond to practice. My aim is to show those terms of opposite sign that disappear in the theory of Lexis-Bortkiewicz.

Not wanting an argument about what one ought to compute to divert us from my main goal, I am computing on this page in the simplest manner in accordance with the customary (but not correct) formula, bringing in only a small correction to the scheme of Lexis-Bortkiewicz: the variation in the probability within each group.

The notation was indicated earlier. I hope that everything here is clear and simple.

Yours,

A. MARKOV, 20 November 1910

Frankly speaking, I doubt that there exists an equality which one should call the Bienaymé-Bortkiewicz formula; in essence all of these quantities are very simple.

No. 21

(Postcard from Markov to Chuprov)

20 November 1910

Your last letter was evidently written hastily. From my subsequent letters you can see clearly that statistics in theory is one thing, and in practice another. I am not obliged to make mistakes in it. On the contrary, I must point out the error. However, read my letters of yesterday with the detailed calculations.

Yours,

A. MARKOV

No. 22

(Letter from Chuprov to Markov)

20 November 1910

Highly esteemed Andrei Andreevich:

The formulas for the mathematical expectation of $\Sigma(\xi_i - \xi)^2$ and $\Sigma(\xi_i - p)^2$ in your letters which I received last evening and this morning are very interesting and inviting, both for their extreme generality and simplicity and for the Einsicht opened up by them into the play of circumstances increasing and reducing the dispersion as against the norm (in the sense of Lexis). The tendency toward an increase caused by the differences among the p_i and the decreasing effect of the differences among the $p_i^{(j)}$ for given i is caught by them with striking clarity. The formulas that generally appeal to statisticians are easily obtained from yours, for example the formula that I introduced on page 350 and the one on page 282[1]. It is very instructive that we can, with different success, approach these cases, both from your formulas and Bohlmann's formulas. This shows the conventionality of that distinction one usually makes between the scheme of independent events with varying probability and the scheme of dependent events, a point of view to which general logical reasoning also leads me.

As to the question of whether to compute the mathematical expectation of $\Sigma(\xi_i - \xi)^2$ or of $\Sigma(\xi_i - p)^2$, I look at it this way. The point is what goal one is pursuing. If one has in view a theoretical analysis of the conditions leading to more or less stability, then it is natural to proceed from $\Sigma(\xi_i - p)^2$. If, however, one has in mind the direct utilization of the given trial, then of course one must lean on $\Sigma(\xi_i - \xi)^2$. In practice, however, as your formulas in the completely general form show clearly, the disparity almost reduces to the usual recipe of dividing the sums by $\mu - 1$ instead of μ.

Sincerely yours,

A. CHUPROV

---

[1]Chuprov, A. A., *Essays on the Theory of Statistics*, M., Gosstatizdat, 1959, page 231.

No. 23

(Postcard from Markov to Chuprov)

21 November 1910

Highly respected Alexander Alexandrovich:

I hope that you will appreciate the significance of my computations although they do not correspond to your schemes in view of the fact that they break with deep-rooted teaching, according to which supernormal stability for independent trials is impossible.

Yours,

A. MARKOV

No. 24

(Letter from Chuprov to Markov)

21 November 1910

Highly esteemed Andrei Andreevich:

I dare to hope that I would not have overlooked the great significance of your formulas, even if they were contrary to my views. One has to break with established constructions when it is required in connection with the success of scientific thought. I am not accustomed to being afraid; the enormous interest of the formulas is obvious. In the field of dispersion theory I think this represents the greatest possible generality in the formulation of the problem and the most striking clarity of solution. It seems to be impossible to go any further in this direction. On the basis of your formula (also bringing in Bohlmann's formula for a parallel clarification from a somewhat different point of view) it is now possible to cover the whole theory of stability in a completely finished form, to sketch its entire construction freely without the yoke of the fortuitousness of its historical growth.

But I still believe that you are bringing us statisticians not a sword but peace. Your results are not a reversal but rather the completion of that which time has done for us. Your attacks on statisticians are concentrated on three points:

1. the error in introducing $\Sigma(\xi_i - p)^2$ instead of $\Sigma(\xi_i - \xi)^2$;

2. the possibility of supernormal stability with independence (the case of equality of all the p_i with inequality of the $p_i^{(j)}$);

3. a concrete example which gives supernormal stability in those cases where statisticians would have expected supernormal dispersion.

Today you put aside the first point as less important; I shall also put it off for the present, all the more as I indicated briefly yesterday how I personally look at the matter. I have considered the concrete example on the enclosed page. As for the second point, you describe the scheme concretely and are convinced that one can formulate it in a way which shows the redundancy of dependence. The case that I considered in my book on pages 281–284—the case "of the average probability of an invariable structure," according to Bortkiewicz's terminology, is the most characteristic example. This case can be made concrete—and perhaps must be if we are not to introduce a direct volitional influence on the course of the trials—in this way: there are n urns; from them one is chosen at random, a drawing is carried out, and the urn is put aside; then another urn is chosen at random from those remaining, etc., until the whole row of urns is exhausted. Dependence here appears to be of the same type as in the removal of a ball from an urn without returning the ball to the urn.

Yours sincerely,

A. CHUPROV

No. 25

(Letter from Markov to Chuprov)

21 November 1910

Highly respected Alexander Alexandrovich:

To tell the truth, I do not like the scheme with double probability[1] which you have indicated.

It is appropriate to introduce such probability in the study of the determination of probability based on observations, but there it is called for by necessity since we must distinguish between the probability corresponding to our data and the probability based on more complete data.

But in the question you have considered I do not see the necessity and I do not understand the motive for introducing it.

I cannot agree with your opinion about the conditionality of the distinction between the scheme of independent events with changing probability and the scheme of dependent events.

The given case presents the peculiarity that the trials are independent, given the magnitude of the probability. Thanks to this peculiarity the computations can be reduced to the case of independent variables. In fact, the mathematical expectation of the sum you considered obviously breaks up into terms of the form

$$E (x_1 + x_2 + \cdots + x_k - kc_0)^2,$$

where $c_0 = E x_1 = E x_2 = \cdots = E x_k$.

By considering this mathematical expectation we easily reduce everything to independent variables since $x_1, x_2, \ldots, x_k$ do not depend on each other, given the appropriate common probability.

Taking this into account we obtain:

$$E (x_1 + x_2 + \cdots + x_k - kc_0)^2 = \Sigma g_\lambda w_\lambda,$$

where $w_\lambda = E (x_1 + x_2 + \cdots + x_k - kc_0)^2$ in the case where

$$E x_1 = E x_2 = \cdots = E x_k = c_\lambda = p_\lambda.$$

Conforming to this we have:

$$
\begin{aligned}
w_\lambda &= E (x_1 - c_\lambda + x_2 - c_\lambda + \cdots + x_k - c_\lambda + kc_\lambda - kc_0)^2 \\
&= E (x_1 - c_\lambda + x_2 - c_\lambda + \cdots + x_k - c_\lambda)^2 + k^2 (c_\lambda - c_0)^2 \\
&= \text{(under the conditions indicated)} = k(c_\lambda - c_\lambda^2) + k^2(c_\lambda - c_0)^2
\end{aligned}
$$

and

$$E(x_1 + x_2 + \cdots + x_k - kc_0)^2 = k\Sigma g_\lambda(c_\lambda - c_\lambda^2) + k^2\Sigma g_\lambda(c_\lambda - c_0)^2.$$

And since

$$\Sigma g_\lambda \cdot c_\lambda = c_0, \quad \Sigma g_\lambda = 1 \quad \text{and} \quad \Sigma g_\lambda \cdot c_\lambda^2 = \Sigma g_\lambda \cdot c_0^2 + \Sigma g_\lambda \cdot (c_\lambda - c_0)^2,$$

we finally obtain a result that agrees with yours.

Allow me to pose a question: has anyone attempted to apply a similar scheme to real data or does it serve only as an example?

If an attempt at applying the scheme has really been made, then I would like to know which elements of the scheme one tried to compute from the observations. This interests me keenly; if the scheme introduced a series of numbers that did not come from the observations, then it cannot, in my belief, have any meaning.

Apropos of this, I consider it necessary to confess that I did not know about one important theorem of Bortkiewicz which you mentioned only in a note, namely that the mathematical expectation of $(1 - Q^2)^2$ tends to zero as the number of observations increases if the probability is constant. On the strength of this theorem, values of Q remote from 1 become unlikely with a large number of observations.

I believe that this theorem, to which little attention is paid, is highly important although it does not quite distinguish values smaller than and greater than one. At the same time I suppose that Q represents only an auxiliary element for research on statistics and can have meaning only for Quetelet's scheme[2].

From your book I gather, perhaps erroneously, that some statisticians are interested solely in the value of Q.

I beg you to accept my esteem and complete respect.

A. MARKOV

Incidentally, can you tell me where to find Bohlmann? I should like to send him my article in which, in my opinion, the general theorem of Bernoulli is extended to many cases of dependent variables.

Yours,

MARKOV

---

[1] By double probability Markov apparently means that, first, one selects an urn by chance (the probability of which is g_i), and then one draws a ball from it at random (the probability of which is c_i) (Editor's note).

[2] By Quetelet's scheme Markov presumably means the scheme of Bernoulli, believing that Poisson's scheme does not differ essentially from Bernoulli's scheme (Editor's note).

No. 26

(Letter from Chuprov to Markov)

22 November 1910

Highly esteemed Andrei Andreevich:

I am sorry that I do not have Bohlmann's address. Having exchanged teaching for private employment (in an insurance company), Bohlmann moved to Berlin where he lives, I recall, in Wilmersdorf, but I have forgotten his exact address. I intend to go abroad at the conclusion of my lectures; on the way through Berlin I shall make inquiries about Bohl-mann's address since I want to send him something, and I shall very will-ingly take what you want delivered to Bohlmann. If this is agreeable with you, send me what you want him to have; I assure you it will be delivered.

No one has applied the scheme that interests you to real data, as far as I know. In general statisticians at present have essentially not gone in for further computations of Q and of the equally valuable comparison of the actual grouping of the frequencies around the mean with the theoretical distribution (assuming constant probability and independence). And the truth is, a large part of the work in the field of study of stability is carried out by mechanical people who are computers rather than researchers. That which has been done by statisticians in this area has been considered in my book, I think with almost exhaustive completeness. That is not much, it goes without saying. But considering that work in this direction has begun very recently, and indeed even now there are few among stat-isticians who are capable of any intelligent and conscientious handling of it, do not judge us too harshly. Indeed it is gratifying that the idea has come to some movement at least.

On the question concerning the conditional nature of the difference between the scheme of independent events with varying probability and the scheme of dependent events, for the present I am not resolved to insist on my own point of view. My reflections on its basis are not complete, and now there is a great deal to think about in connection with your formulas. Perhaps you will take a day or two from your current work to understand all that follows from them. Now it is necessary to think concentratedly on the replanning of the theory of dispersion. Perhaps also it will ultimately be necessary to renounce the views that I expressed to you some days ago under the first impression from your communication. For example, I retract partly what I wrote at that time about the recipe for the division of the sum by $\mu - 1$ instead of μ. Your formulas, of course, do not justify it as much as carry it to its limits.

Sincerely yours,

A. Chuprov

No. 27

(Letter from Markov to Chuprov)

22 November 1910

Highly respected Alexander Alexandrovich:

I do not bring either peace or a sword to statisticians but in the nature of my work I try to understand objectively the aspect of statisticians' inferences that pertains to the theory of probability.

In this connection my attention has been attracted especially by those results which apparently are acknowledged to be impossible, as I judge from your book and the review of it by Bortkiewicz in the *Journal of the Ministry of Public Education.*

To me it seems important to make it clear that these cases are impossible only when Quetelet's scheme in the narrow sense is preserved, i.e., when the cases of Lexis and Bortkiewicz are impossible.

I do not propose any kind of definite scheme, but *I consider what can happen when Quetelet's scheme is violated, i.e., when the probability is varied.* And I find that in addition to the results of Lexis and Bortkiewicz, results of another character are possible.

The case where the probabilities vary within each group, preserving a constant mean, seems extreme to me. Of course I do not consider it normal, but it is also difficult for me to acknowledge as normal the other extreme case where the probability is kept constant within each group and varies only in the transition from group to group. Between these two extreme cases an infinite number of other cases with a deviation in one direction (Bortkiewicz, Lexis) or the opposite are possible.

Here of course it is important to emphasize that I considered independent trials exclusively because I knew the inferences of Lexis and Bortkiewicz referred only to independent trials. Therefore I cannot agree with your scheme which moves from case of independent trials to the case of dependent ones.

In general I regard the double probability unsympathetically and I believe that it is necessary to avoid it as far as possible.

The order of the urns is given without randomness as you also have to assume in the scheme of Lexis-Bortkiewicz if you want to consider the case of independent trials.

If you reduce the Lexis-Bortkiewicz case to the case of dependent trials, then their assertion turns out to be meaningless.

In closing I must confess that the problem of statisticians in their scheme of Quetelet is, for me, completely unclear.

With complete respect,

A. MARKOV

No. 28

(Letter from Markov to Chuprov)

23 November 1910

Highly respected Alexander Alexandrovich:

You can observe from my book *The Calculus of Probability*[1] that I am concerned only with questions of pure analysis that arise under the condition of known basic premises. In the second edition this is noted especially in the preface.

Of course it is very interesting and important that questions of mathematical analysis should be found to be in definite correspondence with the requirements of practice.

However, there cannot be complete coincidence between requirements of practice and questions of mathematical analysis. In connection with this I indicated in the first edition of my book that all experimental sciences are concerned with an approximate search for numbers which do not exist in a strict mathematical sense. Of course I also repeat this in the second edition; I repeat only in the preface that in applications to the study of nature the question of error obtains a special character, for these investigations have to do with quantities that are not fully defined. In one of Bortkiewicz's articles mentioned by you I found a comparison of statistical inferences concerning various physical constants.

Bortkiewicz believes that these constants are some kind of completely defined numbers. However, I know perfectly well that this is false and that every constant in physics represents only some average number pertaining to a certain series of observations.

Of course I do not insist that one can apply the theory of probability to the statistical questions we are talking about.

In general I try not to assert anything except the results of my own computations.

In my book I refer to the question of the applicability of probability theory with indifference, noting only what kinds of assumptions one must make in this connection and expressing doubt about their realization.

And I note that it is completely impossible to prove the validity of that kind of application.

For those who do not agree with *the basic position* of Queteletism about *the existence of probabilities* as numbers other than 0 and 1 in each particular case, it is impossible to prove anything.

Of course if one denies this essential position of Queteletism it is impossible to speak about the applications of probability theory.

But, on the other hand, I should note that it is impossible also to disprove this basic position of Queteletism by any kind of computations.

Of course to this first position there is still a second to be added—about the constancy of this probability.

However, I am convinced that Quetelet himself has hardly insisted on this constancy; I believe that even he can find indications that various probabiltiies can change, depending on the changing conditions of life.

The assumption of constancy of a probability represents only the first, simplest hypothesis. Any research in experimental science begins with a similar hypothesis.

It is quite possible that in a given case it is impossible to go further from the hypothesis although significant fluctuations in the probability would be found.

Certainly one can include all possible assumptions in general mathematical formulas but it is difficult to express the hypothesis of a change in the probability in a form capable of experimental verification. When there is no better hypothesis in the sense of compatibility with observations one has to be content with the first hypothesis. All computations of the quantity Q can show that some fluctuations occur in a number which under the hypothesis is assumed to be constant. But in no way do they change the fundamental position of Queteletism about the existence of probability as a number. The scheme with double probability that you apparently like very much should be called double Queteletism and can in no way be used for the destruction of simple Queteletism.

For one who simply does not acknowledge probability, double probability will hardly be more admissible. What kind of probability is this where one definition of a urn is taken for several defined experiments?

And then again probability appears in ... (the word is indecipherable—Kh. O.). Therefore, it is necessary to define both of these probabilities separately: what meaning do they have without reference to the urn but in nature? And as for numerical results, they cannot provide anything more than simple Queteletism.

I go back to the question about the impossibility of supernormal stability for independent trials. It is quite amazing that Bortkiewicz arrived at his conclusion about such an impossibility but you, thanks to your schemes, do away with supernormal dispersion too when you cited the example[2], known to Bortkiewicz, on page 282 where stability is supernormal for independent trials.

Should this not explain the respect for Lexis who first proclaimed a similar impossibility?[3]

With complete respect,

A. MARKOV

[1]See: Markov, A. A., *The Calculus of Probability*, 4th edit., M., 1924.

[2]Chuprov, A. A., *Essays on the Theory of Statistics*, M., Gosstatizdat, 1959.

[3]W. Lexis actually denied the possibility of supernormal stability if it was not called for by straightforward requirements of law or custom (Lexis, W., *Abhandlungen zur Theorie der Bevölkerungs und Moralstatistik*, Jena, 1903, p. 180). Bortkiewicz also pointed out the condition ("average probability of a constant structure") when supernormal stability can arise in laws of probability theory. On page 282, *Essays on the Theory of Statistics*, Chuprov introduced a formula concerning just this case (Editor's note).

No. 29

(Letter from Chuprov to Markov)

24 November 1910

Highly esteemed Andrei Andreevich:

Apropos of that point of view of Bortkiewicz you mention (contrasting statistical and physical constants), we had much disagreement with him. One must consider that Bortkiewicz did not study the natural sciences; you see he is a lawyer by education.

As to the uses of probability in statistics, the problems for which statisticians turn to the theory of probability are quite varied and not at all confined to that one which you have primarily in view—the approximation of "constants" (in the conventional sense) on the basis of available empirical data. As important if not more important is the question of whether one and the same "constant" lies at the basis of two comparable empirical numbers; in other words (for the simplest case), whether or not the difference between two comparable frequencies goes outside the limit of that which would have been reasonably expected under the assumption of invariability of the probability and of independence of the trials. I think this problem is not necessarily unsolvable, even though at first it seems hopeless.

I say the same about a problem which the fourth essay of my book expresses most clearly: to explain how fluctuations in statistical numbers can depend on time. Here the theory of probability permits us not only to single out sharply the influence of the number of observations, but also to outline more or less systematically a whole class of influences capable of raising or lowering the fluctuations, and, moreover, on the basis of our observations we can often be convinced immediately that these effects are present. For these reasons the general formal character of the conditions giving rise to comparatively high and comparatively low stability is needed, especially the theory of dispersion in probabilistic schemes. The significance of one or another scheme for statistics is closely connected with the possibility of noticing in the relations studied by statisticians the presence of such concurrences of circumstances that more or less correspond to the construction of these schemes. In this sense I consider all the schemes introduced in my book a substantial forward step against Quetelet. And the difficulty, perhaps even the impossibility, of making these results concrete numerically does not in my opinion deprive them of interest.

Yours faithfully,

A. CHUPROV

No. 30

(Letter from Markov to Chuprov)

24 November 1910

Highly respected Alexander Alexandrovich:

At last I have obtained a final clarification of the truth and am convinced that the inferences of Lexis's school with which it has endeavoured to defeat Queteletism are, to a significant degree, caused by inadequate attention to the expressions that are considered.

Up until now all of my attention has been concentrated on the numerator of Q. However, in the expression for Q there is also a denominator that is determined from the observations. Thus, the true value of Q which the computers have to be concerned with all the time has a fractional form

$$\frac{\sum \left\{ \xi_i - \frac{\xi_1 + \xi_2 + \cdots + \xi_\mu}{\mu} \right\}^2}{\frac{\mu}{n} \left\{ \frac{\xi_1 + \xi_2 + \cdots + \xi_\mu}{\mu} - \left(\frac{\xi_1 + \xi_2 + \cdots + \xi_\mu}{\mu} \right)^2 \right\}}$$

$$\left(\text{here } \xi_i = \frac{x_i' + x_i'' + \cdots + x_i^{(n)}}{n} \right),$$

and no one has proved that its mathematical expectation is equal to one. In view of the fact that this expression has a fractional form it is certainly possible to assert that its mathematical expectation does not exactly equal one. Of course one can make some inferences about it, considering the numerator and denominator separately.

But it is hardly possible to prove that within the limits of observation the observed values of Q are unlikely.

With complete respect,

A. MARKOV

No. 31

(Letter from Markov to Chuprov)

24 November 1910

My dear Alexander Alexandrovich:

My last letter, sent just now, was based on a defect. Formally, of course, everything in it is true but if the substance of the matter is gone into, its ending destroys its beginning.

In view of what is known about the arithmetic mean one can, with probability as close to one as desired, which can in fact be computed, establish that

$$\frac{\xi_1 + \xi_2 + \cdots + \xi_\mu}{\mu} = U$$

differs from p by less than a given arbitrarily small quantity.

Hence, of course, one can also draw a conclusion about the ratio

$$\frac{U - U^2}{p - p^2}.$$

Therefore, although calculation of the mathematical expectation of the fraction is difficult and certainly this mathematical expectation is not equal to one, nevertheless it is possible to conclude that the probability of significant deviations of the fraction from one is small if only one can establish the corresponding conclusion about the numerator of the fraction. And for the last purpose the mathematical expectation of the square of the difference, considered by Bortkiewicz, can serve.

With complete respect,

A. MARKOV

I always hasten to admit my own mistakes first.

No. 32

(Letter from Markov to Chuprov)

25 November 1910

Highly respected Alexander Alexandrovich:

Of course one could have nothing against the schemes expounded by you if:

1. It were shown clearly that all of these schemes were inseparably connected with the simplest scheme of Quetelet and could not exist without it.
2. The scheme of variable probability, established in the theory of probability, were not forgotten.
3. The lack of clarity about the notion of independent trials, established in the theory of probability, had not been introduced.
4. These schemes had not led to an erroneous conclusion about the impossibility of supernormal stability for independent trials.

Schemes leading to erroneous results certainly cannot be the object of scientific analysis.

With complete respect,

A. MARKOV

No. 33

(Letter from Markov to Chuprov)

27 November 1910

Highly respected Alexander Alexandrovich:

The more I try to grasp the sense of what you have written the less I agree with you. Right now for example, on page 269 of your book I find: "The intersection of independent sequences of events underlies the law of large numbers."[1]

Here, following Nekrasov, you pay attention to that circumstance without which the law of large numbers can perfectly well survive. It is interesting to observe that, further on, you dwell on the scheme of Poisson which is related to dependent trials.

In my book[2] in the article on independent trials the scheme could not find a place but it is contained or, it is better to say, can be inferred from that material discussed in the article "Extension of the law of large numbers . . . . ." at the beginning of §2 (1907).

In that same year another article was also published in the *Proceedings of the Academy of Sciences.* Thus, independence is not an essential condition but you have omitted one that is essential: *constancy of the probability,* of course determined (for example, 1/10, 1/100, . . . , 99/100, etc.). Restriction of freedom of arbitrariness is also contained in it.

With complete respect,

A. MARKOV

------

[1]Here Markov is right that Chuprov refers incorrectly to the concept of A. Cournot concerning independent causal sequences. This concept proclaims a condition for the origin of "randomness" and is compatible with any form of dispersion since side by side with "independent causal" sequences "interdependent" ones are also quite thinkable; the first insure "randomness" and the second—regulate dispersion. Therefore, it is impossible to associate Cournot's concept so closely with the law of large numbers (Editor's note).

[2]Markov, A. A., *Seclected Works.* Theory of Numbers. Theory of Probability., M., Press of the Academy of Sciences of the U.S.S.R., 1951 (Editor's note).

No. 34

(Letter from Chuprov to Markov)

28 November 1910

Highly esteemed Andrei Andreevich:

The place on page 269 of my book[1] to which you refer has no relation to the point of view you are discussing. That contraction of independent sequences which we are discussing here is observed equally, both when the scheme of independent trials is carried out and when we deal with dependent trials. This is the general logical foundation of the very concept of probability insofar as it claims a definite, objective meaning, and in some way or other is associated with frequencies. However one constructs the scheme, the moments determining the movement of the hand will be thought independent of those that determine the arrangement of the balls, and so on. If all the related sequences are introduced in all their mutual relationships, we shall have to deal, for example, with an open urn from which a person who can see wants to remove a white ball, and there will no longer be any probabilities before us.

I acknowledge fully the connection between the schemes I have set forth and that one which you apparently call the simplest scheme of Quetelet, as well as the logical prius of the latter; all schemes appeal in the end to the law of large numbers in its very simplest form. However, this does not eliminate their differences in those respects which eliminate their role in the contemporary statistical criticism of Queteletism. Strictly speaking, this criticism does not have any points of contact with the criticism of Nekrasov who does not understand—indeed, I think does not know—the real status of the question in the statistical literature.

I shall not dwell on the question of the logical interrelation and the comparative statistical value (as applied to the problem of stability) of the schemes of dependent trials and independent trials (with variable probability). It would lead too far into the field of logic. And also from the point of view of what particularly interests the statistician, it is not very important. If one repudiates the idea of logically reducing one group of schemes to another and defining normal dispersion consistently so that it holds for independent trials and constant probability, the results essential for statistics are not affected. The possibility of normal dispersion outside of these conditions does not trouble the statistician. The impossibility of other than normal dispersion under observance of both conditions is obvious. Only a small retouching of certain formulations would be needed in order that, with a shift of the logical base from the scheme of dependent trials to that which seems more appropriate to you, all basic propositions of the contemporary theory of stability of mass phenomena remain valid.

As for the proposition that the mathematical expectation of Q is equal to one, it, of course, rests on a series of simplifications, in part very rough. First of all: the mathematical expectation of Q^2 is, in fact, properly taken to be approximately equal to one. (In connection with this Bortkiewicz, leaning on the relation $E\{(1 - Q^2)^2\} = 2/\mu$, obtained for Q the expression

$$1 - \frac{1}{4\mu} + \frac{1}{12\mu^2},$$

where μ is the number of frequencies on which Q is computed.) Furthermore, statisticians do not forget that the mathematical expectation of a fraction is being computed when equating the mathematical expectation of Q^2 to one; however, they limit themselves to a reference to the comparative smallness of the fluctuations of the denominator without a more detailed analysis. Finally, the result given by Bortkiewicz for the mathematical expectation of $\{(1 - Q^2)^2\}$ is based on the substitution of the quantities $\frac{\Sigma(\xi_i - p)^2}{\mu}$ and $\frac{\Sigma(\xi_i - \xi)^2}{\mu - 1}$ for each other. All of this is understandably somewhat primitive and a more refined analysis would be desirable, but for the requirements of statistical work one can manage with a rather rough sketch of the mathematical base.

Sincerely yours,

A. CHUPROV

---

[1]Chuprov, A. A., *Essays on the Theory of Statistics*, M., Gosstatizdat, 1959 (Editor's note).

No. 35

(Letter from Markov to Chuprov)

28 November 1910

My dear Alexander Alexandrovich:

I consider it my duty to express my opinion decisively in the press in the *Journal of the Ministry of Public Education* opposing the attitude established in your book[1] toward the basic propositions of the calculus of probability and its theorems.

Only the question of whether, after our correspondence, the second edition will retain the same content as the first stops me.

I most humbly beg you to give me an answer to this question.

By the way, apropos of the scheme cited on page 276 I observe:

1. That all of the schemes with urns serve only as an illustration of the notion of probability as a number, just as a square traced on paper illustrates a geometric square.

2. That the scheme on page 276 illustrates the well known idea about a change in the probability with a change in the data. When we are given which urn has been selected, the probability will correspond to that urn. But while only a certain probability is associated with each urn, the probability of an event will be defined by the theorems on the addition and multiplication of probabilities. After that nothing remains except Bernoulli's theorem.

With complete respect,

A. MARKOV

[1]Chuprov, A. A., *Essays on the Theory of Statistics,* M., Gosstatizdat, 1959 (Editor's note).

No. 36

(Letter from Chuprov to Markov)

29 November 1910

Highly esteemed Andrei Andreevich:

Except for the appendices the second edition of my book was ready for the press when our correspondence began and, to my regret, our exchange of ideas could not be reflected in it. The publication of the book has been delayed somewhat, but I firmly expect to receive copies this week and of course you will get one of the first copies. I deeply regret that an unfortunate accident—the negligence of the errand boy who confused the copies and left you the one intended for A. Kohn—deprived me of the possibility of meeting your objections in good time. Although I do not agree with many things, I cannot fail to recognize your critique of the theory of stability I have set forth as the most profound and principled of all that I have had the occasion to listen to up to this time. I shall await your article with great interest, confident that whatever may be the ultimate results of our discussion, science will be the winner.

Faithfully yours,

A. CHUPROV

No. 37

(Postcard from Markov to Chuprov)

30 November 1910

Highly respected Alexander Alexandrovich:

I most humbly beg you not to be in a hurry to send me the second edition inasmuch as my review is almost finished on the basis of the first edition.

The analysis of the second edition can provide me with new difficulties.

Surely in the second edition you retained the ideas of the first: about probability, about independence and about the law of large numbers.

With complete respect,

A. MARKOV

No. 38

(Postcard from Markov to Chuprov)

1 December 1910

Highly respected Alexander Alexandrovich:

I most humbly beg you to let me know whether the theorem on the impossibility of supernormal stability has been included in the new edition of Czuber's book. If it has, indicate the page.

With complete respect,

A. MARKOV

No. 39

(Letter from Chuprov to Markov)

2 December 1910

Highly esteemed Andrei Andreevich:

Unfortunately I do not have the first edition of Czuber and it is not possible to get it quickly, so I cannot tell you very precisely in what ways the two editions differ and in what ways they agree. I cannot find the example to which you refer. Generally speaking, although Czuber developed the section on the theory of stability significantly in the second edition, he kept all of the former points of view, right up to those that are obviously erroneous (for example, the incorrect interpretation of the formula for the random drawing of balls from an urn without replacement). He speaks about the impossibility of supernormal stability for "unconnected" phenomena in section 205 of the second volume, pp. 46–47.

In order to have a more precise reference, perhaps you will let me know where in the first edition the question that interests you arises, or if you wish, I shall send the second to you.

Faithfully yours,

A. CHUPROV

No. 40

(Letter from Markov to Chuprov)

2 December 1910

Highly respected Alexander Alexandrovich:

Finally my note is finished and sent off. I have expended much effort on it, endeavoring to write in order to convince even you. In order to avoid all polemics I confined myself exclusively to that area where I feel a firm footing.

If I had wished to polemicize I would have dwelt on many points of your argument.

I note them for you personally:

1. there are no independent sequences;
2. the law of universal gravitation does not always hold because two particles with the same electric charge repel each other;
3. the argument about the Napierian logarithm of 3 and the decimal system requires a clarification of the method of proof of the dependence or independence of the number on the system.

I note all this, not for polemics but in order to clarify that I had another objective in mind: to establish the correct view of the calculus of probability and its terms as well as to clear up certain misunderstandings. I can regard the first edition of the *Essays* with indifference but the erroneous reference to Bohlmann in the second must grieve[1] me.

With complete respect,

A. MARKOV

---

[1] Of course I shall not begin to polemicize on this.

No. 41

(Letter from Markov to Chuprov)

3 December 1910

Highly respected Alexander Alexandrovich:

Thank you very much for the reference you sent me. You have sent me exactly what I needed. I do not have the second edition of Czuber's book and so I was not sure whether Czuber had corrected his mistakes himself. You have not understood my example. It is not in Czuber's book nor could it be. I am sorry that I made you look for it in Czuber's book; I thought that you yourself had guessed the thing that I meant to say so I indicated my thought by a single example. Of course it is my fault that I did not give a proper explanation but there was no space on the postcard. By my example Lexis's case of supernormal dispersion is inseparably linked to the opposite case of supernormal stability. Here is my example. I hope that now I have presented it clearly.

The series is characterized by the following probabilities:

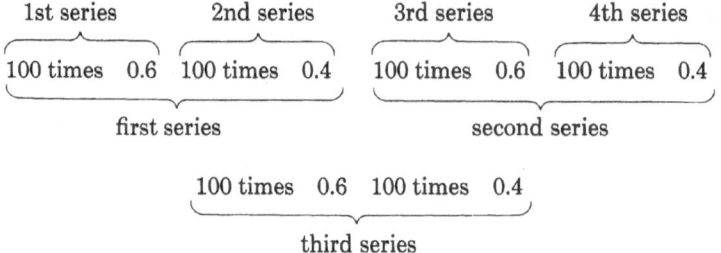

When series with 100 terms are considered we have Lexis's case but when we take them in groups of 200 we obtain the opposite case:

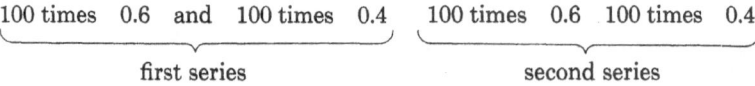

With complete respect,

A. MARKOV

If this is not clear, please let me know.

No. 42

(Letter from Markov to Chuprov)

4 December 1910

Highly respected Alexander Alexandrovich:

Thank you for sending the second edition of your book[1] because now all fear that my note is belated is removed. I have not examined what changes you have introduced in other sections but here I find only the addition of some references.

All that I am going to do is to mention that the second edition has appeared. I have already said that, having good cause to expect (or to think or to assume) that such an incorrect attitude (toward the fundamental propositions of the calculus of probability and toward the law of large numbers) can spread.

Here I now add in a footnote that your book is out in a second edition.

I apologize in advance that for definiteness I have associated many things with the *Essays,* although similar things are expressed in other places as well.

To your professor, about whom I am not saying a word, a note probably will seem even to be wrong. And Bortkiewicz probably will not like my reference to *your* theorem of Poisson.

After sending the references to you, I sent a letter to Czuber. The words, "the profound mathematical research of Pearson," in the new edition astonish me.

With complete respect,

A. MARKOV

---

[1]Chuprov, A. A., *Essays on the Theory of Statistics,* 2nd edition, St. Petersburg, 1910 (Editor's note).

No. 43

(Letter from Markov to Chuprov)

5 December 1910

Highly respected Alexander Alexandrovich:

I most humbly beg you, if it won't be too much trouble, to pass on the two articles I have sent you to Bohlmann when you are abroad. One of them, as you can see, appeared not in the works of the Academy of Sciences or of the Kazan Mathematical Society *which you did not know about,* but in the foreign journal *Acta Mathematica.*

Regarding the inscription written in the book, I will say that you have nothing to regret. If you had valued my remarks fully, they would have prevented you from publishing the second edition of the book.

Your argument with the Moscow school is amusing to me because it was established long ago that independence is not a necessary condition, but that another condition is necessary in order to show the real facts— the *constancy* of the probability[1].

With complete respect,

A. MARKOV

---

[1]This constitutes the aim of my work.

No. 44

(Letter from Markov to Chuprov)

7 December 1910

Highly respected Alexander Alexandrovich:

In connection with your last letter I must remark that I am not so flippant as you assume. I shall not go a step out of that region where my competence is beyond any doubt.

I do not intend to start any argument with you about the formulation

of any kinds of questions. I have not expressed any fundamentally negative attitude toward views accepted in statistics nor do I intend to do so.

As for supernormal and subnormal stability in connection with independence *in the sense of the calculus of probability,* I hope that my
example was sufficiently convincing for you.

In one of your previous letters you wrote that it was clear to you and
now I have to conclude that you did not understand it because you are
not convinced by it of the possibility of subnormal dispersion when independence is preserved. Where, finally, is the truth?

Did you understand it or did you not?

With complete respect,

A. Markov

No. 45

(Letter from Markov to Chuprov)

24 November 1912

My dear Alexander Alexandrovich:

E. E. Slutsky's book[1], *Theory of Correlation,* and the preparation of
the third edition of my *The Calculus of Probability* have induced me to
turn my attention again to the questions with which our correspondence
was concerned. E. E. Slutsky's book interests me but does not attract me.
I have heard that you have recommended it to students, from which I
conclude that you found it worthy of attention.

In the new edition of *The Calculus of Probability* I intend to dwell on
the application of the method of least squares to the determination of
probability with application even to a numerical example taken from page
188 of E. E. Slutsky's book.

I find this application very useful for clarifying the method. Permit me
to mention the basic points in a few words so that the formula I want to
ask you about will be completely clear.

Let p be the unknown probability and q its complement so that $p + q = 1$. Let N series of observations be performed with n experiments in
each with the approximate equalities

$$p \approx \frac{m_1}{n}, p \approx \frac{m_2}{n}, \ldots, p \approx \frac{m_N}{n}.$$

From them we infer the new

$$p \approx \frac{m_1 + m_2 + \cdots + m_N}{n \cdot N} = p^0.$$

The question concerns the estimation of the errors, in other words the determination of the mathematical squares of these errors. In this estimate the number n remains fixed but the m_i are not determined (0, 1, 2, ...). The mathematical expectation of the square of the error of each equality $p \approx m_i/n$ is equal to $\dfrac{p(1 - p)}{n} = \dfrac{k}{n}$, and that of the final equality $(p \approx p^0)$ is equal to $\dfrac{p(1 - p)}{n \cdot N}$.

For the unknown k we have two approximate equalities, free from systematic error.

The first

$$k \approx \frac{n \cdot N}{n \cdot N - 1} \cdot p^0(1 - p^0),$$

is usually replaced by $k \approx p^0(1 - p^0)$.

The second is

$$k \approx \frac{1}{N - 1} \sum n \cdot \left\{ \frac{m_i}{n} - \frac{m_1 + m_2 + \cdots + m_N}{n \cdot N} \right\}^2.$$

All these formulas are known. But now I shall go on to one that is less well known. It concerns bounds for the approximate equalities mentioned above.

Apparently Bortkiewicz was the first to consider this question. Right now I cannot remember where his results were published and how they were derived although I do recall the following; I spoke with him sometime about them and found them essentially correct. I am presenting here my exact results so that they can be compared with those of Bortkiewicz. I shall take only the second approximate formula, for the first follows from it after dividing up the whole series into the separate experiments ($n \equiv 1, N \equiv Nn$). The expression that we shall consider is

$$W = \frac{1}{N - 1} \sum \left\{ \frac{\mu_i}{n} - \frac{\mu_1 + \mu_2 + \cdots + \mu_N}{n \cdot N} \right\}^2$$

$$= \sum \frac{\mu_i^2}{n \cdot N} - \frac{2}{n^2 \cdot (N - 1) \cdot N} \sum \mu_i \cdot \mu_j,$$

where $\mu_i = m_i - np$. When W is written in this way all the computations are especially simple because the mathematical expectation of $\mu_i = 0$. The mathematical expectation of $\mu_i^2 = npq$.

In addition we need only the mathematical expectation of μ_i^4 for which the formula

$$E \, \mu_i^4 = 3n(n - 1)p^2 \cdot q^2 + npq(p^3 + q^3)$$
$$= 3n^2p^2q^2 + npq(1 - 6pq).$$

can be established. It is also possible to write the mathematical expectation of μ_i^3 but it is not necessary since μ_i^3 is multiplied by μ_j and the mathematical expectation of $\mu_j = 0$ (in this lies the advantage of the equality indicated).

It is now apparent that the mathematical expectation of W is equal to pq, therefore we write $pq \approx W$; but no one has proved and no one will prove that the mathematical expectation of $\sqrt{W} = \sqrt{pq}$.

Of course, from the approximate equality $pq \approx W$ it is possible to infer that $\sqrt{pq} \approx \sqrt{W}$, but there is no appropriate method for estimating the error of the last equality.

Meanwhile, E. E. Slutsky's purpose is served apparently by formula (11) which, in my opinion, is also incorrect, as is (12)². The incorrectness of formula (12) is shown by his result on page 104, stemming from his confusing the deviations from the mean with infinitesimally small quantities. I am interested in the correct formula for estimating the error of the equality $pq \approx W$.

The mathematical expectation of this mean-squared error is equal to[3]

$$E \, W = pq.$$

Then

$$E \, W^2 = \frac{1}{n^2 \cdot N^2} \sum E \, \mu_i^4 + \frac{2}{n^2 N^2} \sum E \, \mu_i^2 \cdot \mu_j^2$$
$$+ \frac{4}{n^2 N^2 (N - 1)^2} \sum E \, \mu_i^2 \cdot \mu_j^2$$

and furthermore

$$E \, W^2 - p^2 \cdot q^2 = \frac{1}{n \cdot N} \{3np^2q^2 + (1 - 6pq)pq\} + \frac{1}{N}(N - 1)p^2q^2$$
$$+ \frac{2}{N(N - 1)} p^2q^2 - p^2q^2 = \frac{2}{N - 1} p^2q^2 + \frac{1 - 6pq}{nN} pq.$$

This formula is exact and from it, for large n and N, follows the approximation

$$E \, (W - pq)^2 = \frac{2}{N} \, p^2 q^2.$$

Comparing it with the formula indicated by E. E. Slutsky I see a big resemblance but I have the number 2 in the numerator.

Since my formula is very simple I am interested in knowing whether or not it was given by Bortkiewicz. Surely he obtained something similar. By the way, this formula shows that it is most advantageous, from the point of view of precision of the derivation, to replace N by the single value $n \cdot N$, i.e., to look at the equality

$$k \approx \frac{n \cdot N}{n \cdot N - 1} \cdot p^0 (1 - p^0).$$

According to the general formula the mathematical expecation of the square of its error is equal to

$$\frac{2}{N \cdot n - 1} \cdot p^2 \cdot q^2 + \frac{1 - 6pq}{n \cdot N} \, p \cdot q,$$

or approximately $\dfrac{1 - 4pq}{n \cdot N} \, p \cdot q$. Finally, I conjecture how E. E. Slutsky obtained his formula (11). Probably everything stems from confusing the deviations from the mathematical expectation with infinitesimals.

Such a deduction is no good at all.

I apologize for taking your time. However, I hope that the subject of my letter is not lacking in interest for you. Incidentally, permit me to ask whether it is worth my while to acquaint myself with Leontovich's *Elementary Textbook on the Application of the Methods of Gauss and Pearson?*

Please be assured of my complete respect.

Yours,

A. MARKOV

---

[1]On page 188 of his book *Theory of Correlation and Elements of the Study of Distribution Curves*, E. E. Slutsky cites Weldon's experiment with throwing 12 dice and counting the number of occurrences of 5 or 6 pips (Editor's note).

E. E. Slutsky (1880–1948)—prominent Russian mathematician, statistician and economist (Editor's note).

²The formulas given by E. E. Slutsky are:

$$E_\sigma = 0.67449 \, \frac{\sigma}{\sqrt{2N}} \ldots, \tag{11}$$

$$E_\nu = 0.67449 \, \frac{\nu}{\sqrt{2N}} \, \sqrt{1 + 2 \left(\frac{\nu}{100}\right)^2} \ldots, \tag{12}$$

where σ is the quadratic deviation, $\nu = \sigma/M$ is the coefficient of variation, M is the mean, N is the number of observations.

These formulas are approximate and correspond to the level of development of the theory of statistics at that time. As is clear from the text of his postcard of 25 November 1912, A. A. Markov later changed his opinion about formulas (11) and (12), acknowledging that they "can be obtained" in the case of a normal distribution (Editor's note).

³Translators' note: a literal translation of the apparently trivially incorrect Russian text is: The mathematical expectation of the square of this error is equal to

$$E \, W^2 = p^2 q^2.$$

No. 46

(Postcard from Markov to Chuprov)

25 November 1912

Highly respected Alexander Alexandrovich:

My doubt about the formula for the probable error (E. E. Slutsky's formula (11)) for the Gaussian case has been dispelled. I have convinced myself that it can be obtained.

With complete respect,

A. MARKOV

No. 47

(Letter from Markov to Chuprov)

26 November 1912

My dear Alexander Alexandrovich:

Thank you for sending the series of articles by Bortkiewicz from which it is obvious that the difference between my calculations and Bortkiewicz's is not great. It has been explained in earlier letters. I also acknowledge that Bortkiewicz's calculations merit attention. I shall give my calculations in the new edition of my book and, of course, I shall point out that the first person who carried them out successfully was Bortkiewicz.

I hope that he did not have any predecessors. Assuming the Gaussian law, the approximate (limiting) result is more easily obtained. But I think that before Bortkiewicz nobody at all made similar computations even with the help of Gauss' formula.

Montemartini's article, apparently, has no relation to this question. But the reference in it to the book of Yule, *Theory of Statistics* attracted my attention. Probably from Yule's book I could become acquainted with Pearson's work, if I knew English. However, everything that I know about it from accounts of his work by others does not predispose me to it.

For example, in §197 and 198 of Czuber's book an utterly impossible derivation of Pearson's formulas is given, contradicting them.

I assume that the derivation was given by Pearson; there is no formal error in it but in the final result it was forgotten that in the limit it is necessary to equate a certain variable to zero (see page 24)..

Have you not come across an indication anywhere that this inference is essentially incorrect, and if it were corrected would inevitably lead to the same Gaussian law? You will do me a great favor if you will give me an answer to this question. I am corresponding with E. E. Slutsky in connection with his book which interests me although I do not like it very much. Even from it I want to take one example (§32), but I shall not consider it in Pearson's way.

Sincerely yours,

A. Markov

No. 48

(Letter from Chuprov to Markov)

28 November 1912

Highly esteemed Andrei Andreevich:

I don't recall that I have come across anywhere that objection which you make in connection with the exposition of Pearson's work by Czuber. But I must say that the derivation given by Pearson himself and his school runs quite differently. There is no discussion of urns. An interpolation problem is set up in a straightforward general form: given a series of frequencies, find the equation of a curve representing them reasonably well, starting from two properties common to the majority of the curves of frequencies that are actually observed: 1) the curve at one of its ends touches the axis of the abscissa; 2) at some distance from its end the curve has a maximum. On this basis the equation

$$\frac{dy}{dx} = \frac{y(x + a)}{F(x)}$$

is taken as a starting point. Then $F(x)$ is expanded in a MacLaurin series and integration is carried out under the assumption that the series can be stopped at the first, the second, or the third term; one does not go beyond the third term. The method of moments serves to determine the constant from the empirical data.

Thus, from their construction Pearson's formulas tend toward the category of "empirical" curves—true, with a slight touch of a "rational" element in the initial data, distinguishing them somewhat, from a logical point of view, from the ordinary parabola. The justification for Pearson's construction is that the statistical material is actually very often represented rather well by his formulas. Mortara is not concerned with the questions which you are now studying but does touch on the theme we discussed in our correspondence in the past. Therefore I decided to send it to you also when I learned that you are preparing a new edition of *The Calculus of Probability.*

Yule is now the most interesting textbook in theoretical statistics. But he does not go along with Pearson in everything. W. Palin Elderton's book, *Frequency-Curves and Correlation,* is the most suitable one for familiarizing oneself with Pearson's theories.

Sincerely yours,

A. CHUPROV

No. 49

(Letter from Markov to Chuprov)

1 December 1912

Highly respected Alexander Alexandrovich:

On Thursday I prepared a letter to you but did not send it because after a conversation with Mr. Chekanovsky I had to change its contents.

With regard to my inquiry, it turns out that the derivation given by Czuber is taken directly from Pearson. However, Mr. Chekanovsky who apparently is well acquainted with the literature informed me that the groundlessness of Pearson's derivation had already been observed by K. E. Ranke and Greiner in the article "Das Fehlergesetz und seine Verall- gemeinerung durch Fechner und Pearson" and that Pearson answered them in his *Biometrika,* London, 1910, vol. IV, but what the reply was he did not know. P. A. Nekrasov mentions the article of Ranke and Greiner, but one of two things is true: either their observation does not agree with mine, or P. A. Nekrasov did not understand it since he does not say directly that they pointed out the groundlessness of this result.

Chekanovsky appeared like a meteor or a comet and then vanished; he promised to bring me both the article by Ranke and Greiner and volume IV of *Biometrika* but as of now he has not done so.

On the statement of the question you formulated in your letter and about which Mr. Chekanovsky also spoke to me, the formulas already fall outside the field of probability theory and become more or less successful empirical formulas.

Of course I do not deny the usefulness of empirical formulas but I do not concern myself with them. The other question disturbs me greatly. Up until now I have associated the name of Bortkiewicz with the question of bounding the error of an approximate expression by using the squared error.

However, Mr. Chekanovsky thinks that a similar bound was made much earlier than Bortkiewicz's work with the help of Gauss' formula. He was unable to indicate by whom and when this was done. Apparently Bortkiewicz himself did not refer to this either. It seems to me that before Bortkiewicz no one had found it necessary to bound such an error, one might say, of the second order.

Perhaps, though, I have missed something. It disturbs me greatly. I hope that Mr. Chekanovsky shows up because he took the little book of E. E. Slutsky from me. If he has disappeared once and for all from my horizon, then perhaps I shall ask you to lend me, for a short while, volume

IV of *Biometrika*, if it contains something for me. Until then, however, I do not need *Biometrika*.

Sincerely yours,

A. MARKOV

In Bertrand's chapter X, §192, there is a computation similar to Bortkiewicz's.

No. 50

(Letter from Markov to Chuprov)

2 December 1912

Highly respected Alexander Alexandrovich:

It turns out that I have been cruelly mistaken regarding the significance of Bortkiewicz's work. Fortunately this error was expressed in print in only one phrase.

I imagined that he was the first to study the estimate of the error in the determination of the probabilities of deviations. It turns out that the main results were obtained long before his work.

My error is explained by the fact that before the correspondence with you I was not interested in these questions, not attaching any importance to them.

As for his law of small numbers, I don't deny Poisson's formula

$$\frac{m^x e^{-x}}{1 \cdot 2 \cdots m}$$

but I do not understand what Bortkiewicz's discovery is.

Judging by the notice of his "Zur Verteidigung des Gesetzes des kleinen Zahlen," I assume that Bortkiewicz's discovery which he called the law of small numbers states that for small numbers the coefficient Q is close to one where it ought not to be (*Jahrbücher für Nationalökonomie und St.,* 1910, Bd 39).

I have deduced this from the controversy with Gini which surprised me where he apparently computed one and the same number but obtained two very different results:

1.037 and 3270.

Thus, he destroys Gini but, in my opinion, rather reveals the incorrectness of his own computations.

We see that the numbers in Table No. 13

6, 8, 7, 5, 14, 8, 9, 4, 8

have to do with quite different numbers of observations (people)

42542, ... , 65531, . . . , 66909.

From this point of view of the theory of probability one would expect to reduce all of the numbers of the first table to the same number of observations, otherwise all of these numbers have completely different meanings. Is it possible, from the point of view of statistics, to combine in a series the first number corresponding to 42,542 observations and the number 8 corresponding to 66,909 observations?

If the number n in all the groups were the same, then we would have to have $Q_{n-x} = Q_x$.

It seems to me that in putting forth Q Bortkiewicz puts on the first plane something that ought to be on the second and does not deal adequately with his observations.

The arguments with different people are based on formulas of an excessively special character which do not have an absolute meaning. However, I cannot go into a detailed study of these questions. If I am mistaken perhaps you will point out my error to me?

Sincerely yours,

A. MARKOV

No. 51

(Letter from Chuprov to Markov)

3 December 1912

Highly esteemed Andrei Andreevich:

We do not differ very much in our estimation of the "law of small numbers": I do not see an important discovery here either; I think that it attracted attention in larger measure by its striking name than by its substance. Furthermore, up to now in spite of Bortkiewicz's additional questions in a verbal discussion I cannot determine with certainty how Bortkiewicz himself wishes to see the content of the law, which of the four alternatives formulated in the footnote to page 398 of the second edition of my *Essays*[1] he chooses.

For me the significance of Bortkiewicz's work in this area lies more in certain results that are obtained incidentally than in the law of small numbers itself.

As regards Bortkiewicz's work in the theory of errors, here the statement of the question of fluctuations in Q is objectively original; but subjectively, Bortkiewicz is also original in that he was not familiar with the work of his predecessors when he wrote these articles. It was the same with the correction to the trapezoidal formula proposed by him in 1890: Bortkiewicz's formula was also used earlier in the statistical literature but Bortkiewicz had not run across it even very recently; only last year, when he again claimed his copyright in a lecture to the International Statistical Institute, I pointed out to him that before him this formula was given by the Swiss Schertlin and still earlier by Becker.

Sincerely yours,

A. CHUPROV

---

[1]Here A. A. Chuprov has in mind the following place in *Essays on the Theory of Statistics:* "Bortkiewicz himself in "Gesetz der kleinen Zahlen" gives an insufficiently clear definition: it remains unclear what should properly be called the law of small numbers. There are four alternatives to choose from:

1. the relation of Q to n, made known by Lexis and extended by Bortkiewicz to the case of small probabilities;
2. the mathematical formula proposed by Poisson as the distribution law in the case of probabilities so small that Laplace's formula does not apply to them;
3. the fact ascertained by Bortkiewicz of the agreement of the actual distribution of "small" numbers with this formula; the greater closeness of the variations to the normal level in the case of unlikely phenomena than in the case of phenomena whose probability is great—in other words, the relation of Q to p.

All of these statements can lay claim to the name "law of small numbers" since small numbers enter into them as an essential element of their content with the only difference that in 2) and 3) we are talking about absolutely small numbers of occurrences with large numbers of trials, in 1) the discussion concerns small numbers of trials with any number of occurrences, and in 4) it concerns a small number of occurrences with any number of trials." (Editor's note).

Translators' note: The fourth alternative was not given in the editor's footnote.

No. 52[1]

(Letter from Markov to Chuprov)

4 December 1912

Highly respected Alexander Alexandrovich:

In your letter of December the third you mention some kind of correction to the trapezoidal rule found by Bortkiewicz. Not long ago I had in hand an article by Bortkiewicz which was apparently devoted to this correction. I did not pay much attention to it and now I cannot find it (the article). However, I remember that Bortkiewicz refers to me. From your words I conclude that I am to blame after all, having acknowledged the correction as new. Now I do not want to make inquiries; however, I think that I did not state this emphatically. In any case I didn't make inquiries at that time and I didn't attach great importance to the formula. For me it was important only to establish that Bortkiewicz understood the matter and his work was not meaningless.

Yours,

A. MARKOV

If Monday is inconvenient for you name another day. You can speak to me by the telephone which has been installed in my apartment.

---

[1]Up to letter No. 67 of A. A. Markov, A. A. Chuprov's replies, which we did not succeed in finding in the archives, are absent (Editor's note).

No. 53

(Letter from Markov to Chuprov)

8 December 1912

Highly respected Alexander Alexandrovich:

On the basis of your letter I plan to send the messenger from the Academy of Sciences to you on Monday between three and four o'clock for *Biometrika*. However, I am afraid to take all of it up to the last volume.

I assume that the first five volumes are quite enough for me since in *The Calculus of Probability* I shall have to speak about Pearson. But, besides *Biometrika,* I would like to obtain, if you have it, *Archiv für Anthropologie Neue Folge,* Band IV which contains the article by Ranke and Greiner "Das Fehlergesetz ..." I shall no longer dwell (at length) on the applications of the theory of probability to statistics. About the coefficient Q, I shall merely remark that in the numerical example I took from Pearson it is close to one. Our previous correspondence had some influence on the first chapter of my book[1]: I dwell more on the clarification of the fundamental concepts but, of course, I do not change the direction I have taken.

By the way, everywhere possible I exclude the completely undefined expressions "random" and "at random"; where it is necessary to use them I introduce an explanation corresponding to the particular case.

Sincerely yours,

A. MARKOV

My telephone number is 616-20

---

[1]Markov, A. A., *The Calculus of Probability,* 4th edition, Moscow, 1924 (Editor's note).

No. 54

(Letter from Markov to Chuprov)

15 January 1913

Highly respected Alexander Alexandrovich:

Thank you and the library of the Polytechnic Institute very much for the books which I am returning. I did not get anything from them but it was important to find out that it was not necessary for me to get anything from them. Now permit me to make two requests of you.

Firstly, do you know: the year 1913 is the two hundredth anniversary of the law of large numbers (*Ars Conjectandi,* 1713), and don't you think that this anniversary should be commemorated in some way or other? Personally, I propose to put out a new edition of my book, substantially expanded. But I am raising the question about a general celebration with the participation of as large a number of people and institutions as possible.

The second question concerns an original statistical investigation which I have carried out and with which I propose to conclude my book.

The character of the investigation, which embraces a sequence of 20,000 letters, is shown in the example below. Here I have 200 almost independent variables each of which represents the result of 100 dependent trials. The arithmetic mean of the 200 numbers is 43.2 and the probability that a letter be a vowel is 0.432. Computing the sum of squares of the deviations of the numbers 42, 46, ... from the mean 43.2, we obtain the sum 1022.8 which, divided by 200 and then by 100, gives 5.114 and 0.05114.

1	2	3	4	5	6	7	8	9	10		1	2	3	4	5	6	7	8	9	10
м	о	й	д	я	д	я	с	а	м		у	к	а	н	о	б	о	ж	е	м
ы	х	ч	е	с	т	н	ы	х	п		о	й	к	а	к	а	я	с	к	у
р	а	в	и	л	к	о	г	д	а		к	а	с	б	о	л	н	ы	м	с
н	е	в	ш	у	т	к	у	з	а		и	д	е	т	и	д	е	н	и	н
н	е	м	о	г	о	н	у	в	а		о	ч	н	е	о	т	х	о	д	я
ж	а	т	с	е	б	я	з	а	с		н	и	ш	а	г	у	п	р	о	ч
т	а	в	и	л	и	л	у	ч	ш		к	а	к	о	е	н	и	з	к	о
е	в	ы	д	у	м	а	т	н	е		е	к	о	в	а	р	с	т	в	о
м	о	г	е	г	о	п	р	и	м		п	о	л	у	ж	и	в	о	г	о
е	р	д	р	у	г	и	м	н	а		з	а	б	а	в	л	я	т	е	м
3	7	2	5	5	3	5	4	3	5		5	6	3	6	6	3	5	3	4	5

$$(1,6) = 3 + 3 = 6 \qquad\qquad (1,6) = 5 + 3 = 8$$
$$(2,7) = 7 + 5 = 12 \qquad\qquad (2,7) = 6 + 5 = 11$$
$$(3,8) = 2 + 4 = 6 \qquad\qquad (3,8) = 3 + 3 = 6$$
$$(4,9) = 5 + 3 = 8 \qquad\qquad (4,9) = 6 + 4 = 10$$
$$(5,10) = 5 + 5 = 10 \qquad\qquad (5,10) = 6 + 5 = 11$$

	42	

	46	

Count for 500			Number of vowels among		
(1,6)	(2,7)	(3,8)	(4,9)	(5,10)	100 letters
6	12	6	8	10	42
8	11	6	10	11	46
11	7	6	11	5	40
11	7	7	9	10	44
13	15	13	4	8	43
49	42	38	42	44	215
Number of vowels in the modified hundreds					Number of vowels in 500 letters in sequence

The square of the coefficient of dispersion turns out to be

$$\frac{0.05114}{0.432 \cdot 0.568} \approx 0.21.$$

I consider this number with the help of my formula

$$\frac{1 + \delta}{1 - \delta}.$$

I find the conditional probability that a letter be a vowel given that the preceding letter was a vowel and the conditional probability that a letter be a vowel given that the preceding letter was a consonant: δ denotes their difference,

$$\delta = 0.128 - 0.663 = -0.535,$$

and therefore

$$\frac{1 + \delta}{1 - \delta} = \frac{465}{1535} \approx 0.3.$$

Because the sums are independent there are only small changes in the sum of squares of the deviations when the observations are combined by twos, by fours and by fives: here are the numbers:

1022.8	827	975.2	1004
original	by twos	by fours	by fives

With a different grouping of the letters in hundreds, which I have indicated in the above example, only 40 independent numbers are obtained, all 200 by fives. The sum of squares of the deviations is 5788.8. The square of the coefficient of dispersion is $\dfrac{0.28994}{0.432 \cdot 0.568} = 1.18$ which corresponds to independence of the trials grouped in hundreds. But the sums are dependent, because of which, under grouping by twos, by fours and by fives, quite different results are obtained for the sum of squares of the deviations:

$$3551.6 \quad 3089.2 \quad 1004;$$

the last number is nearly six times smaller than the original 5788.8.

To me this example seems instructive in many respects. I hope that no one has considered such an example up until now. But of course one cannot be fully confident; therefore, please let me know whether or not you have come across a similar example anywhere. If you have, please tell me where so that I can compare the results.

Sincerely yours,

A. MARKOV

No. 55

(Letter from Markov to Chuprov)

27 January 1913

Highly respected Alexander Alexandrovich:

The question of the necessity of marking the 200th anniversary of the law of large numbers has been raised in the Academy of Sciences and a commission was organized to discuss what we can do. The commission will meet in a day or two at my place.

The discussion concerns the translation into Russian of *Ars Conjectandi* or at least the 4th chapter which even now has not lost interest. However, I should like to organize a conference with the participation of representatives of statistics, including those who are not members of the Academy of Sciences.

Of course ideas about the law of large numbers in the contemporary view of it will vary but it can hardly be disputed that the theorem of Jacob Bernoulli is the basis of it.

One of the people on whom I can rely, perhaps even the only person, is you. Therefore, I turn to you with the question: what do you think of this idea? Further, the participation of different institutions in the form of their representatives seems possible to me.

Faithfully yours,

A. MARKOV

No. 56

(Letter from Markov to Chuprov)

31 January 1913

Highly respected Alexander Alexandrovich:

Your plan—to publish a special collection of articles—requires thorough discussion and dealings with many persons. The organization of an honorary celebration on an international level lends great importance to it, of course, but at the same time introduces considerable complications.

For me it is even unclear whom it would be appropriate to involve in collaboration in the collection. It is clear that in the current year it is impossible to carry out your plan. The initiative for carrying it out should belong to you; I am prepared to assist you, but under the condition that it does not interfere with carrying out the simple celebration which I proposed yesterday to the Academy's commission and which they approved.

In the autumn of this year we intend to organize a general conference devoted to the law of large numbers, in which we count on your participation. Besides you and me, it was proposed to bring in Professor A. V. Vasiliev to this work.

Then it was proposed to translate only the fourth chapter of *Ars Conjectandi;* the translation will be done by the mathematician Ya. V. Uspensky who knows the Latin language well, and it should appear in 1913.

Finally, I propose to do a French translation of the supplementary articles in my book with a short foreword about the law of large numbers, as a publication of the Academy of Sciences.

All of this should be scheduled for 1913 and a portrait of J. Bernoulli will be attached to all the publications.

In connection with your idea about attracting foreign scholars, I cannot fail to note that the French mathematicians, following the example of Bertrand, do not wish to know what the theorem of Jacob Bernoulli is. In Bertrand's *Calcul des Probabilités* the fourth chapter is entitled "Théorème de Jacques Bernoulli," the fifth is entitled "Démonstration élémentaire du théorème de Bernoulli." But neither a strict formulation of the theorem nor a real proof of it is given.

The theorem is found in Poincaré but it is not clearly connected with the name of Jacob Bernoulli. Too slight a respect for the theorem of J. Bernoulli is also observed among many Germans. It has reached the point that a certain Charlier has written an article "Ueber die Strenge Form der B. Theorems" in which he showed a complete lack of familiarity with the theorem.

Sincerely yours,

A. MARKOV

No. 57

(Letter from Markov to Chuprov)

4 February 1913

Highly respected Alexander Alexandrovich:

I have not dwelt on the question of the importance of the law of large numbers to physics. It would be appropriate to talk with Prince Golitsyn[1] about this. If he wishes to speak about that, it will be very good. It seems to me it is awkward to appeal to other physicists. I suppose that we shall not restrict the lengths of the speeches beforehand but I, confining myself to theorems, do not propose to speak a long time. Whereupon, permit me to return to the question of small numbers. It seems to me that here we are concerned with facts of a purely arithmetical character which prove exactly nothing.

The phenomena which are revealed clearly with extensive numbers are concealed with small numbers. Small numbers, frankly, play a trick on me, giving me a result the opposite of what is expected.

When I counted the number of vowels all together the dependence between the letters was clearly revealed by the large change in the probability. When I began to count the number of appearances of a single letter a, the probability of its appearance became small (0.0717) and could no longer vary significantly: the numerical value of the number δ which I introduced did not reach 0.08 and my coefficient of dispersion turned out to be larger than 0.92/1.08.

The practical coefficient of dispersion (according to you it would still be appropriate to take the square root) turned out to be $\frac{6562}{6656} \approx 0.98$, very close to one. And this coefficient did not increase as was the case in the count of all of the vowels, but on the contrary decreased a little when distant letters were combined into groups. With a new distribution of letters, by permitting a succession of two and three a's, I obtained for the coefficient of dispersion the number $\frac{6438}{6656} \approx 0.96$.

I cannot recognize this result as legitimate; on the contrary I am convinced that real legitimacy was not shown because of its weakness and turned out to be concealed by chance circumstances, having acquired a substantial value with small probability.

Sincerely yours,

A. Markov

---

¹Golitsyn, Boris Borisovich (1862–1916)—prominent Russian physicist, academician, founder of seismology (Editor's note).

No. 58

(Letter from Markov to Chuprov)

2 March 1913

Highly respected Alexander Alexandrovich:

Today the Academy of Sciences accepted my proposal for a commemoration of the anniversary of the law of large numbers.

Now before summer we ought to agree on the contemplated speeches.

It would be very good if you could meet with me, together with A. V. Vasiliev for this purpose. Of course you live fairly far from me but perhaps you will find some time not too inconvenient for you?

I am usually at home and I shall discuss things with A. V. Vasiliev who has very many different duties, after you indicate the times that are more convenient or less convenient for you. There are five or six weeks left at our disposal before Easter. Such a meeting is not necessary, of course, but it would be useful; therefore I would like to arrange it if it turns out to be possible.

My telephone is 616-20.

Sincerely yours,

A. MARKOV

No. 59

(Letter from Markov to Chuprov)

8 March 1913

Highly respected Alexander Alexandrovich:

A. V. Vasiliev has just informed me that next week he can come to my home in the evening, both on Monday and on Tuesday. And so I most humbly ask you to choose one of these days for our meeting.

As to the time, I hope that 8:00 o'clock in the evening will not be an inconvenient hour for you.

So I most humbly ask you to let me know on what day and hour I can expect to see you. The telephone can serve this purpose: here is my telephone number: 616-20.

Sincerely yours,

A. MARKOV

A. V. Vasiliev prefers not 8:00 o'clock but 8:30 in the evening, but I am afraid that this will be late for you.

No. 60

(Letter from Markov to Chuprov)

31 October 1913

Highly respected Alexander Alexandrovich:

The meeting devoted to the law of large numbers is planned for 1 December at 2:00 p.m. I hope that you will not find this time inconvenient for yourself. The fact is that many find the 24th of November inconvenient and on the 21st of November another meeting has long since been scheduled. It was proposed to arrange our meeting in the evening (8:00 o'clock) on the 21st of November but I find the evening hour unsuitable. I assume that it is also more convenient for you to come to St. Petersburg in the day and not late in the evening. Then, S. F. Oldenburg[1] asks us to announce the titles of the lectures so that the program can be printed beforehand. And so I ask you to write me how you want to entitle your lecture. Mine will be called: "The law of large numbers (or an essay on the development of the law of large numbers) as a collection of mathematical theorems." According to this title I certainly will not touch on the application of the law of large numbers but to make up for it I shall also include, under the law of large numbers, the expression for the probability by the well-known integral.

Of course I shall mention Poisson since the expression "law of large numbers" is his, but I shall not attach great importance to him.

If this name has a particularly significant meaning for statisticians, then you will speak about it.

The order of the talks is naturally the following: A. V. Vasiliev, then I and finally you.

Tel: 616–20.

With sincere devotion,

A. MARKOV

---

[1]S. F. Oldenburg (1863–1934)—prominent Russian scientific-orientalist, academician. In 1904–1929 he was the permanent secretary of the Academy of Sciences (Editor's note).

No. 61

(Letter from Markov to Chuprov)

4 November 1913

Highly respected Alexander Alexandrovich:

Please make a selection from the three titles you have indicated or think up a new one. It is desirable to settle on the titles this week since our press prints very slowly. Regarding your talk, I assume that it will probably be printed in the publications of the Academy in a more extensive form than will be given orally, of course only in Russian. A translation can be published in another place. By the way, allow me to ask you: was there some conversation about the 200th anniversary of the law of large numbers at the meeting of the International Statistical Institute?

Sincerely yours,

A. MARKOV

No. 62

(Letter from Markov to Chuprov)

3 December 1913

Highly respected Alexander Alexandrovich:

In order to avoid any misunderstanding I hasten to inform you that I cannot subscribe to A. V. Vasiliev's project to put together a general collection of the three talks. We joined together only temporarily on the basis of mutual independence. From the outside the celebration appears to have turned out fairly successfully. But there is not complete solidarity between us. In your talk statistics stood first and foremost, and applications of the law of large numbers were advanced that seem questionable to me.

By subscribing to them I can only weaken that which for me is particularly dear: the rigor of judgments I permit.

For example, it is very important for me to observe that Poisson did not prove his theorem; moreover, I cannot consider a statement a theorem unless it is established precisely.

Your talk harmonized beautifully with A. V. Vasiliev's talk but in no way with mine. I had to give my talk since the 200th year of a mathematical theorem was being celebrated, but I do not intend to publish it and I do not wish to. The combination of your talk with A. V. Vasiliev's talk without mine is quite possible, as is clear, by the way, from P. Yushkevich's article in the newspaper *The Day* (of 2 December, No. 327); and I suggest it be done.

Sincerely yours,

A. Markov

No. 63

(Letter from Markov to Chuprov)

3 February 1915

Highly respected Alexander Alexandrovich:

Do you not have Duncker's book *Die Methode der Variationstatistik* and can you not lend it to me for a while? I found a mention of it in R.

Orzhentskiĭ's work *Textbook of Mathematical Statistics.* Incidentally, permit me to ask you: what do you think of the exercises on pages 81–84 of Orzhentskiĭ's work?

Sincerely yours,

A. Markov

Has your lecture at the Academy of Sciences or a more extensive article on this subject been published anywhere?

No. 64

(Letter from Markov to Chuprov)

29 January 1916

Highly respected Alexander Alexandrovich:

One of the professors at the Polytechnic Institute told me that your students are carrying out many calculations relating not only to lower but also to higher mathematics. These words lead me to believe that you and the students are studying, among other things, Pearson's curves, computing various elements of them and drawing them.

I sincerely ask you to inform me whether my conclusion is right because I am looking for someone who would be interested in studying Pearson's curves and working with them in practice without the religious fervor of Pearson. Apropos of this, I can say that I have made a count of the consonants, dividing them into two groups similar to the earlier ones but the result turned out entirely different. However, I was forced to stop this count due to the blindness with which I am threatened.

Faithfully yours,

A. Markov

No. 65

(Letter from Markov to Chuprov)

1 February 1916

Highly respected Alexander Alexandrovich:

The question about Pearson's curves interests me in the following way: whether fitting them is a well-defined operation and whether they really reproduce the observations well in the examples cited. One such example is cited by Czuber on page 33 of his *Wahrscheinlichkeitsrechnung und ihre Anwendung auf Fehlerausgleichung Statistik und Lebensversicherung,* 2nd edition, vol. 2, Leipzig, 1908, 1910; he takes it from Pearson. My question concerns this example.

Firstly, from the data cited, in my belief it is impossible to determine the location of the center of gravity and all the more impossible to find μ_2 and μ_3, not making deliberately false assumptions. Therefore, it utterly amazes me that in Pearson's equation

$$y = 8882.45 \cdot \left(1 + \frac{x}{2.4573}\right)^{0.36891} \cdot e^{-0.150127x}$$

the exponent γ is determined even to six decimal places. On the other hand, I do not find in the given example either in Czuber or in Pearson a proper comparison of the results given by his curve with the actual numbers. The only thing I see clearly is that between 0 and 1, Pearson's curve (the beginning of it) is not zero and thus it does not at all give what is observed. The verification in other intervals (from 1 to 2, from 2 to 3) requires complicated calculations.

I do not find these calculations in Czuber and do not at all know how the people who use Pearson's methods carry them out. I naturally have the thought that no one thinks of carrying out such calculations. It is important to note that Pearson's curve should reproduce the given numbers, not as the *ordinates* but as the *areas,* the determination of which presents difficulty. So these are the reasons why I asked you whether or not you and your students are studying Pearson's curves.

Sincerely yours,

A. MARKOV

No. 66

(Letter from Markov to Chuprov)

3 March 1916

Highly respected Alexander Alexandrovich:

Once again I turn to you in connection with the "law of small numbers." If by these words the well-known formula of Poisson is implied then it is beyond any doubt. If, however, by these words some kind of special stability of small numbers is meant, then it is impossible to substantiate it in theory and it can be supported in practice only by a few examples. I would be interested in familiarizing myself with these examples. Can't you inform me of them?

As for the examples indicated in v. Bortkiewicz's article "Zur Verteidigung des Gesetzes der kleinen Zahlen," *Jahrbücher f. Nat.-Oek. u. St.*, Bd. XCIV (Dr. F., Bd XXXIX), Jena, 1910, I find them unconvincing. In one example everything is based on a sample of 25 (1st example), in another (§10) clearly heterogeneous data are combined into one whole. I combine the more homogeneous data a) and c) and obtain numbers that are not nearly as good at those of Bortkiewicz:

$$
\begin{array}{cccccc}
0 & 1 & 2 & 3 & 4 & 5 \\
5 & 7 & 4 & 8 & 3 & 1
\end{array}
\quad \frac{56}{28} = 2.
$$

a) and c) are combined because for them an integer is directly obtained that makes it possible to compare our numbers with Bortkiewicz's tables; after dividing all of these numbers by 28 we obtain:

0.178;   0.25;   0.143;   0.286;   0.107;   0.036

instead of the tabled numbers

0.1353;   0.2707;   0.2707;   0.1804;   0.0902;   0.0527.

Bortkiewicz's Table 3 is composed, as it turns out, of completely heterogeneous parts. Bortkiewicz treats each line of the 14 (!) pieces of data separately and then adds them up. In this way his last line is obtained.

If he had taken his number 3.47 as a base and found the corresponding numbers from his basic table (Table 3), then different numbers would have been obtained:

$0.0302 \times 112 \approx 3.38$ $0.0334 \times 112 \approx 3.74$

$$0.1057 \times 112 \approx 11.84 \qquad 0.1135 \times 112 \approx 12.7$$

$$0.1850 \times 112 \approx 2.07 \qquad 0.1929 \times 112 \approx 21.6$$

etc.

The first two numbers for (0) and (1) agree quite poorly with reality. Then, computing for a) and c) the coefficient of dispersion I obtain the number

$$\sqrt{\frac{98}{56}} = \sqrt{\frac{7}{4}} = \sqrt{1.75} = 1.3\ldots,$$

which exceeds Bortkiewicz's number

$$\frac{2.15}{1.86} = 1.1\ldots$$

Here, clearly, the expansion of the field of investigation reduces the coefficient of dispersion.

In my opinion the stuff done by Bortkiewicz is very poor and the agreement of the formulas of the calculus of probability with such poor material is highly questionable.

Did Bortkiewicz continue his Table §10I and others in the following years, which now would probably number 20, or not? I have observations on some small numbers also: the number of appearances of the letter ю among 100 letters in sequence. Here is the result of one tally:

0	1	2	3	4
53	32	9	3	3

and here is the result of another:

0	1	2	3	4
53	23	19	4	1.

The difference in the number of appearances of 1 and 2 is quite considerable. The first result is in good agreement with the Poisson-Bortkiewicz formula with

$$m = (32 + 18 + 9 + 12) \div 100 = 0.71;$$

and the second is in poor agreement with it

$$[m = (23 + 38 + 12 + 4) \div 100 = 0.77].$$

As a rule, according to my observations, curves, even based on a large amount of data, need not necessarily agree too closely with these or other formulas of the calculus of probability. And they have meaning so far as one can transfer them from one set of data to another. The last remark relates to Pearson's curves. Assume that observations on 2000 crabs are carried out. Some sorts of tables are compiled concerning these crabs. Then similar tables are compiled for another 2000 crabs and a comparison of one of these tables with the other is made. That which is common to the two can be meaningful.

It is still better if many tables agree. But it seems to me that no one has made many tables of similar composition and compared them. However, generally in this regard I do not dispose of any data.

So, at the present moment I would like very much to know on what statistical material the "law of small numbers" is based, provided it has not been entirely abandoned by statisticians. The last assumption is refuted by the book by A. Kaufman[1].

Sincerely yours,

A. MARKOV

---

[1]Kaufman, A. A., *Theory of Statistics*, Moscow, 1909 (Editor's note).

No. 67

(Letter from Chuprov to Markov)

4 March 1916

Highly esteemed Andrei Andreevich:

If one calls the very formula of Poisson the law of small numbers it is now in use in various fields of scientific work, in particular in physics where a number of researchers have arrived at it, not knowing about either Bortkiewicz or Poisson or even about Abbé who became interested in it in his time in connection with the problem of counting blood corpuscles, work with which physicists would do well to be familiar.

If the name is given to the fact of better conformity of experimental data to theoretical calculations in the case of rare phenomena, this fact has by no means received general recognition. Pearson writes, for example, straightforwardly (in the introduction to Soper's table): a very good coincidence of data with the formula is not often encountered.

On the whole the vagueness that I mentioned in the second edition of the *Essays* (at the end of the footnote to page 398) has recently been pretty well cleared up but not to the advantage of the law of small numbers.

As for A. Kaufman[1], as far as I know no one, even as prominent as he is, is advancing the law of small numbers nor has anyone done so.

At the end of the letter you raise again the question about the logical content of Pearsonian formulas. Apparently our positions are fairly close here. I look at the matter this way. The formula for the distribution curve of the numbers (Pearsonian or some other), as in general any interpolative formula in statistics, finds its justification first of all by reproducing material in a concise form that is otherwise difficult to comprehend. If it reproduces it sufficiently correctly without an excessive number of fitted constants, that alone gives it a certain value as an instrument of scientific work. Moreover, if it turns out that for a certain family of phenomena you constantly encounter a formula of one and the same form only with different constants, the formula acquires additional value: a "nomographic" content is instilled in it, as I would venture to say, and besides that, with the possibility of focusing attention on the constants of the formula we obtain a convenient method for further investigation of the phenomenon. Take even the well-known formula of Pareto for the distribution of the number of people with various levels of income. This formula in very many cases gives the numerical order perfectly well and although the attempts to give it any rational content cannot be considered successful, it is quite helpful in the statistical study of incomes.

Of course a formula gets its richest content when it is possible to interpret it, if only as in the case of the Gaussian curve. Take as an example in statistics the heights of recruits. An investigator observes that the heights of so many thousands of recruits measured by him fit Gauss' curve beautifully; in such a case the curve has for him a descriptive "ideographic" value; it may be used for interpolative purposes (for example, for changing from inches to centimeters, etc.). Furthermore, suppose that wherever the recruits are measured their heights invariably fit Gauss' curve: this form of the curve then becomes typical for the given phenomenon. We know, finally, that Gauss' curve is the curve of errors; in connection with this the mean height has a special meaning for us; we begin to put into this mean a different content than into one about whose mechanism we have no idea.

Pearsonian curves hardly try to reach the third stage. Only in the initial attempt to approach the skewed curve from the hypergeometric series are there traces of such an attempt. The second stage is attainable and in biology (with crabs, etc.) it is perhaps reached.

Sincerely yours,

A. Chuprov

No. 68

(Postcard from Markov to Chuprov)

5 March 1916

Highly respected Alexander Alexandrovich:

I would strongly urge you to suggest to your students that they verify the computation in the notorious work "Das Gesetz ...".[1] In my opinion the computations in this paper are incorrect: on page 23 instead of $[\epsilon''(x)]^2$ = 5.48, I get 7.6; on page 24 instead of 143.1 and 92.1, I get 139 and 97. I fear for my eyes but I can still see.

Yours,

A. MARKOV

---

[1]He means the work of v. Bortkiewicz "Zur Verteidigung des Gesetzes der kleinen Zahlen"—*Jahrbücher f. Nat.-Oek. u. St.*, Bd. XCIV (Dr. F., Bd. XXXIX). Jena, 1910 (Editor's note).

No. 69

(Letter from Markov to Chuprov)

5 March 1916

Highly respected Alexander Alexandrovich:

Permit me to ask you one more question. Has anyone checked Bortkiewicz's computations in "Das Gesetz ..."?

According to my computations the number $[\epsilon_0(x)]^2$ = 4.60 on page 22 and the table on page 24 (not 143 but 139) are incorrect. And then, has anyone clarified the fact that for small numbers the coefficient of dispersion cannot be large, no matter how these numbers are combined?

Yours,

A. MARKOV

No. 70

(Letter from Markov to Chuprov)

7 March 1916

Highly respected Alexander Alexandrovich:

Among the numbers given by Bortkiewicz himself I find such a combination for which, by correct count of the days, Q_x is greater than 3 ($Q_x^2 > 9$).

Take 42 and 52 in the third example and calculate the coefficient of dispersion, taking into account the composition of 42 and 52 ("Zur Verteidigung des Gesetzes der kleinen Zahlen," Jena, 1910, S. 225 and 226).

Yours,

A. MARKOV

No. 71

(Letter from Markov to Chuprov)

8 March 1916

Highly respected Alexander Alexandrovich:

I recognize fully the importance of Poisson's formula and consequently I consider the tables drawn up by Bortkiewicz and Soper quite valuable. But of course I allow it to be applied only in the case of an extensive sequence of homogeneous groups of observations with the same small probability and with an equal number of observations in each group (at least approximately so). But I consider its application to series of 9 groups absurd. Even a series of 99 groups is in itself short if one expects good agreement with probability theory. However, if it is recognized to be heterogeneous and must be broken up into 11, then all of Bortkiewicz's numerical computations are a game with numbers and agreement of the results with that or any other formula can testify only to a skillful selection of numbers. In particular, $\{\epsilon_0''(x)\}^2$, obtained as the mean of 11 numbers, does not correspond to any series of observations. In Bortkiewicz's article, "Zur Verteidigung des Gesetzes der kleinen Zahlen," he brings in a series of very different Q_α^2 from which he deduces the mean. In my opinion, with the admitted heterogeneity of the series, this average has no meaning. Even if the series are homogeneous one must make the computation with all of the observations simultaneously. Then it is necessary to turn one's attention to the number of observations in each group, and as a result the coefficient of dispersion will not turn out to be nearly so small. Out of the 11 (eleven) values of Q_a^2 cited by Bortkiewicz more than a third (4) turn out to be smaller than 1 which contradicts Bortkiewicz's theory but can be explained beautifully from my point of view. At the beginning of the article "Zur Verteidigung des Gesetzes der kleinen Zahlen" Bortkiewicz brings in the formula $E\,(Q_a^2) = 1 + \alpha\{E(Q^2) - 1\}$. In my opinion there is no firm basis for this formula but Bortkiewicz considers it beyond doubt and uses it with $\alpha = 0.001$ and $Q = 5$.

I would suggest to him, rather, to apply this formula in the way that he does to the case

$$Q = 0.88 \quad \text{and} \quad \alpha = 0.01.$$

Man hat also Erwartungs gemäss

$$Q = 10 \cdot \sqrt{-0.2256 + 0.01} = 10 \cdot \sqrt{-0.2156}.$$

My view of the law of small numbers (?) is quite different. We have here simply an arithmetic phenomenon: from small numbers it is difficult to have a large coefficient of dispersion.

The formula that you cite gives a coefficient of dispersion that must be larger than one. Actually for small numbers this coefficient can very often be less than one. If a series consists only of zeros and ones, then the coefficient of dispersion, one can say, according to Bortkiewicz is always less than one, and this circumstance will remain true with a small number of twos.

It is not difficult to see that the coefficient of dispersion cannot be larger than the square root of the largest number. The question can perhaps be treated as probability theory, but another part of it. Several numbers are taken at random, not exceeding a given number; how great is the probability that the coefficient of dispersion computed from these numbers will be larger than a given number and what is the average value of this coefficient?

The complete solution of this question is difficult but I tried computing the coefficient of dispersion for three numbers not exceeding 4 and I found few results exceeding one. I did not carry through completely a similar calculation for four numbers and for three numbers not exceeding 5.

In any case I am convinced that large coefficients of dispersion should be encountered comparatively rarely in arbitrary sequences of small numbers. In my opinion the explanation for Bortkiewicz's notorious law also lies in this. I wish to express such an opinion in the press if nobody has expressed it clearly up to now, and to confirm it by examples with a small number of cases of numbers not exceeding 4.

Regarding interpolatory curves, in my opinion one should distinguish between two cases:

1. when one variable (or both) has only discrete values (0, 1, 2, 3, . . .) and
2. when we have to do with continuous variables. In the second case areas are given and according to them one must select the curve. In the first case the curve is determined by specific values. The second case is far more complicated than the first. Pearson apparently has it in mind. But in this case the moments of different order are not given and cannot serve as a basis for fitting the curve. It is necessary that the areas under the curve correspond as closely as possible to the observed values. The question is fairly tricky.

For Pearson's formula, an examination of it is complicated by the form of the function under the integral sign for which it is appropriate to compute the integral only between certain boundaries. The Laplace-Gauss curve is convenient in that all computations are reduced to one table. It is astonishing to me about the crabs, that their measurements give something completely different than the measurements of the height of recruits. Does this not occur because of insufficiency of data and from the fact that one chases too much after details which cannot be repeated in other observations? However, I am not sure that measurements of the height of recruits always give results that coincide with the well-known curve. Where would the original results of the measurements be found?

Sincerely yours,

A. MARKOV

No. 72

(Letter from Chuprov to Markov)

10 March 1916

Highly esteemed Andrei Andreevich:

The literature on the statistics of heights is fairly extensive. A summary of it is in D. N. Anuchin's work on the geographical distribution of height in Russia; a newer bibliography can be found in Lexis's article "Anthropologie und Anthropometrie" in *Handwörterbuch der Staatswissenschaften*.

The data appropriate for fitting a Gaussian curve are not cited by Anuchin. A wealth of material on Russia can be found, by the way, in a publication of the Central Statistical Committee: "General Military Service in the Russian Empire in the first decade 1874–1883." (*Statisticheskiĭ Vremennik Rossiĭskoĭ Imperii,* Series III, No. 12). In Table III the distribution is given in eleven groups for the provinces for the whole decade and for each year separately; in Table IV the distribution is given in nine groups for the districts. But the groupings are large—from vershok[1] to vershok (except the first).

The comparison of such series with the Gaussian curve is given in Baxter's *Statistics, Medical and Anthropological, of the Provost-Marshal-General's Bureau,* Vol. 5, Washington, 1875.

Edgeworth in one of his earlier works (on statistical methods—in the anniversary volume of the London Statistical Society for 1875), pointing out that not only does the curve retain its form but also that nearly the same measure of accuracy (squared error) is ordinarily obtained, cites a sequence of summary values for different measurements but he does not give the sequences themselves.

The work of Livi is very interesting to a statistician. Livi showed that the finding of Bertillon, much talked of in earlier times, (the two-humped curve of height as evidence of the mixing of races in the département of Doubs), comes from arithmetic negligence: the measurements were made in centimeters and were converted into inches for publication, and they converted them idiotically so that into the average inch grouping, instead of three centimeters there were often two: as a result a saddle was obtained.

Another remark of Livi, very much to the point, is his observation that in measurements one often sees a preference for round numbers, not as striking as in the statistics of height but very clear at times all the same. In the analysis of data it is necessary to pay attention to this situation.

Sincerely yours,

A. CHUPROV

---

[1]One vershok = 1 3/4 inches (Translator's note).

No. 73

(Letter from Markov to Chuprov)

10 March 1916

Highly respected Alexander Alexandrovich:

Thank you very much for the references. I immediately sent to the Academy library for the *Statisticheskiĭ Vestnik*. For my part I can direct your attention to a circumstance I observed in searching for real data referred to in the *Army Statistical Review of the Petersburg Military District*. Cited there are measurements of height: less than 2 ar. 2 ver.[1], from 2 ar. 2 ver. to 2 ar. 2 1/2 ver., from 2 ar. 2 1/2 ver. to 2 ar. 3 ver. and so on in intervals of one vershok. The actual numbers (if not actual, then in any case primary) cited in the *Medical* (army medical) *Journal* give the number of people in intervals of 1/4 vershok and each number refers to a precisely determined height.

The tables in the *Army Statistical Review* are calculated for a total of 1000 people and I am convinced that the numbers pertaining to a, $a + 1/4$, $a + 1/2$ and $a + 3/4$ ver. are combined into one group, forming a group from a to $a + 1$ ver., but in my opinion it is from $a - 1/8$ to $a + 7/8$ ver.

The tables in the *Army Statistical Review* are calculated for a total of 1000 people and I am convinced that the numbers pertaining to a, $a + 1/4$, $a + 1/2$ and $a + 3/4$ ver. are combined into one group, forming a group from a to $a + 1$ ver., but in my opinion it is from $a - 1/8$ to $a + 7/8$ ver.

For this reason I am also beginning to have doubts about the authenticity of the numbers in the *Stat. Vestnik* which are concerned with approximately the same time. Then I am amazed that all of the data are for only the first ten years of general military service.

I treat with skepticism every alteration there which is made without my knowledge and here it is clear to me that the height of recruits was measured with an accuracy of 1/4 vershok while nevertheless there remained the impression of half vershoks and whole vershoks.

Where can one find data for subsequent decades?

In the given case one cannot consider the separate values of the function as given but rather areas. And except for the areas there are no other data. The fitted curve should give the magnitudes of these areas as well as possible. Such a comparison of reality with Gauss' curve has been made; I find it, for example, in *Astronomy* by N. Ya. Tsinger, which prompted me to look for the original source. Nowhere do I find a similar comparison for Pearson's curves but the calculations are based on a form of data which in fact is not available.

With regard to the coefficients of dispersion for small numbers, I made the following computations, changing Bortkiewicz's examples. For suicides of girls I considered, instead of Bortkiewicz's combinations (15, 9, 1) eleven combinations (15, 10, 0, . . . , 15, 0, 10) and always obtained quite a small coefficient. For Bortkiewicz's last example 144, 91, 32, 11, 2, I changed the middle numbers 91, 32, and forming 124 combinations I found for all dispersion coefficients: 1.09; 123.0; 122.1; . . . ; 0.123.

Sincerely yours,

A. MARKOV

---

[1]One arshin = 28 inches, one vershok = 1 3/4 inches (Translator's note).

No. 74

(Letter from Markov to Chuprov)

21 March 1916

Highly respected Alexander Alexandrovich:

Please let me know where one can find the formula

$$E(Q_\alpha^2) = 1 + \alpha\{E(Q^2) - 1\}$$

which Bortkiewicz mentions in the article "Zur Verteidigung des Gesetzes der kleinen Zahlen," calling it Lexis's formula. Apparenly something like it can be found in §14 of Bortkiewicz's article "Das Gesetz der kleinen Zahlen."

However, Bortkiewicz's computations are incorrect for one should have considered not

$$\sum \frac{(p_i' - p_0)^2}{\sigma} \quad \text{but} \quad \sum \frac{(p_i' - p_0')^2}{\sigma},$$

since the values of p_0 are not at our disposal. Therefore, Bortkiewicz's formula in §14 should be recognized as erroneous.[1] Unfortunately I am unable to obtain the issue of *Statisticheskiĭ Vremennik* which you pointed out, either at the university or at the Academy of Sciences.

Sincerely yours,

A. MARKOV

$$E \sum \frac{(p_i' - p_0)^2}{\sigma} = \frac{\sum p_i(1 - p_i)(\sigma - 1)}{\sigma^2 \cdot n}$$

$$= \frac{\sigma - 1}{\sigma} \left\{ \frac{p_0 q_0}{n} - \frac{\sum (p_i - p_0)^2}{\sigma \cdot n} \right\};$$

$$E \sum \frac{(p_i' - p_0)^2}{\sigma} = E \sum \frac{(p_i' - p_i)^2}{\sigma} + \sum \frac{(p_i - p_0)^2}{\sigma}$$

$$= \sum \frac{p_i(1 - p_i)}{\sigma \cdot n} + \sum \frac{(p_i - p_0)^2}{\sigma}$$

$$= \frac{p_0 q_0}{n} + \frac{n - 1}{n} \sum \frac{(p_i - p_0)^2}{\sigma}.$$

Bortkiewicz's formula, as you see, absolutely does not correspond to the facts.

---

[1]With a correct computation under Bortkiewicz's assumption the coefficient of dispersion turns out to be less than one.

No. 75

(Letter from Markov to Chuprov)

22 March 1916

Highly respected Alexander Alexandrovich:

An important error slipped into my computation of yesterday. Although Bortkiewicz's computation also was not entirely correct the mistake is not so great as it appeared to me yesterday. One can consider Bortkiewicz's formula, as I first thought, roughly true when the probability in the series does not change. Otherwise, of course, Bortkiewicz's conclusions fall.

Yours,

A. MARKOV

No. 76

(Letter from Chuprov to Markov)

22 March 1916

Highly esteemed Andrei Andreevich:

The formula

$$E(Q_\alpha^2) = 1 + \alpha\{EQ^2 - 1\}$$

is obtained directly from the approximate relation pointed out by Lexis: $\mu^2 = \mu'^2 + \mu''^2$ (see my *Essays*, 2nd edition, page 385, footnote) if one assumes that the component μ'' is one and the same both for the whole population and for the part of it being considered.

I have already called Bortkiewicz's attention to the correction to formula 6, page 30 in "Das Gesetz der kleinen Zahlen." The disparity of course is not great; in the calculation of the deviations from p_0' I obtain

$$\frac{p_0 q_0}{n} + \frac{n \cdot \dfrac{\sigma}{\sigma - 1} - 1}{n} \sum \frac{(p_i - p_0)^2}{\sigma}.$$

But theoretically the more precise formulas have that valuable advantage (besides their theoretical accuracy) that for further more complicated calculations the work goes more smoothly. So too with the usually accepted inexact formula:

$$E \frac{t(1 - t)}{s} = \frac{p(1 - p)}{s}$$

(p denotes the probability; t the relative frequency in s trials). In practice, with arithmetical work, one can of course use it without the correction. But further theoretical constructions obtain a far more elegant form if we take, as is required

$$E \frac{t(1 - t)}{s} = \frac{p(1 - p)}{s}\left(1 - \frac{1}{s\sigma}\right),$$

and then, properly, it is also necessary to take in the denominator of Q^2:

$$E \frac{t(1 - t)}{s - \dfrac{1}{\sigma}}.$$

A similar inaccuracy is also in Bortkiewicz's formula in the memoir "Ueber die Zeitfolgezufälliger Ereign.", page 33: one should have $\dfrac{2m^2}{\sigma - 1} \cdot \dfrac{s - 1}{s - \dfrac{1}{\sigma}}$ instead of $\dfrac{2m^2}{\sigma} - 1$.

In my personal collection of the press of the Central Statistical Committee there is an issue which would interest you. If you wish I will send it to you. For a much later time there is information given up to 1893 (the same groups as with the Central Statistical Committee but not according to separate provinces and districts, only for Russia as a whole) in the publication *A Century of the Ministry of War, 1802–1902*. General Staff. "An historical essay on recruitment of troops," part III, book I, Section II, page 300, St. Petersburg, 1914.

For the rest apparently one has to turn to the annual report which I have not happened to get. The other day I had occasion to carry through a rigorous deduction of the proposition: $E\ Q^2 = 1$. Following Lexis's example statisticians accept it without any proof for it is just impossible to respect as proof the following argument: the mathematical expectation of the numerator is equal to the mathematical expectation of the denominator; consequently, the mathematical expectation of the fraction is equal to one (as, by all appearances, Lexis reasoned in his time). It is not difficult to prove that $E\ Q^2 = 1$ for the limiting case of an infinitely large number of trials (or an infinitely large number of sequences). But for an arbitrary number of trials and sequences the proof is more complicated. I succeeded in showing that if we let

$$x = \frac{\sum\limits_{i=1}^{\sigma} (t_i - t)^2}{\sigma - 1}, \ y = \frac{t(1 - t)}{s - \dfrac{1}{\sigma}},$$

(t_i is the relative frequency in the ith sequence, t the arithmetic mean, s the number of trials, σ the number of sequences), then for every n: $E\ xy^n = E\ y^{n+1}$, from which it is not difficult to obtain

$$E\frac{x - y}{y} = 0 \quad \text{and} \quad E\left(\frac{x}{y}\right) = 1.$$

In passing I studied the derivation of the general formula for the mathematical expectation of an arbitrary power of the difference of the relative frequency and the probability. I obtain the desired formula in two forms: (z_s is the number of occurrences of an event in s trials):

I. $\displaystyle E(z_s - sp)^{2r} = \sum_{i=0}^{r-1} s^{r-i} \cdot p^r q^r \cdot \sum_{j=0}^{2i} (-1)^j \cdot a_{r+i,2i,j} \cdot \left(\frac{q}{p}\right)^{i-j};$

$$E(z_s - sp)^{2r+1} = \sum_{i=0}^{r-1} s^{r-i} \cdot p^r \cdot q^{r+1} \cdot \sum_{j=0}^{2i+1} (-1)^j$$

$$\cdot \, a_{r+i+1,2i+1,j} \cdot \left(\frac{q}{p}\right)^{i-j}.$$

Here the quantities a are defined by the relations:

$$a_{i,0,0} = 1 \cdot 3 \cdot 5 \ldots (2i - 1);$$
$$a_{i,h,j} = (2i - h - 1) \cdot a_{i-1,h,j} + (i + j - h) \cdot a_{i-1,h-1,j}$$
$$+ (i - j) \cdot a_{i-1,h-1,j-1},$$

whence, among other things

$$a_{i,h,0} = a_{i,h,h}$$

and

$$\sum_{j=0}^{2h+1} (-1)^j \cdot a_{i,2h+1,j} = 0.$$

II. $E(z_s - sp)^{2r} = \sum_{i=0}^{r-1} s^{r-i} \cdot \sum_{j=0}^{i} (-1)^j b_{r+i,i,j} \cdot (pq)^{r-i+j};$

$$E(z_s - sp)^{2r+1} = (q - p) \sum_{i=0}^{r-1} s^{r-i} \cdot \sum_{j=0}^{i} (-1)^j d_{r+i+1,i,j}$$

$$\cdot (pq)^{r-i+j},$$

the quantities b and d are related to the quantities a_i by the relations:

$$b_{i,h,l} = \sum_{j=0}^{l} \{(c_{2h-l-j}^{l-j} + c_{2h-l-j-1}^{l-j-1}) \cdot a_{i,2h,j}\};$$

$$d_{i,h,l} = \sum_{j=0}^{l} c_{2h-l-j}^{l-j} \cdot a_{i,2h+1,j}.$$

I have obtained these formulas in a straightforward way, quite rigorously. In fact, II coincides with the result obtained by Chebyshev's method using an integral. For $E(z_s - sp)^{2r}$ I obtain, for example, the first terms in the form:

$$1 \cdot 3 \cdot 5 \ldots (2r - 1) \cdot s^r \cdot p^r \cdot q^r + 1 \cdot 3 \cdot 5 \ldots (2r - 1)s^{r-1}$$

$$\times q^{r-1} \cdot r(r - 1) \left\{ \frac{2r - 1}{18} - \frac{4r + 1}{9} pq \right\} + 1 \cdot 3 \cdot 5 \ldots (2r - 1)$$

$$\times s^{r-2} \cdot p^{r-2} q^{r-2} \frac{r(r - 1)(r - 2)}{2430} \cdot \{A\},$$

where

$$A = \frac{20r^3 - 60r^2 + 31r + 15}{4} - (40r^3 - 30r^2 - r + 3)pq$$
$$+ (80r^3 + 120r^2 + 7r - 21)p^2q^2.$$

We can obtain the first two if we take the integral in the form in which you computed it in the article "On a problem of Jacob Bernoulli".

Sincerely yours,

A. CHUPROV

No. 77

(Letter from Markov to Chuprov)

23 March 1916

Highly respected Alexander Alexandrovich:

Your proposition $E\,Q^2 = 1$ has greatly excited my curiosity. I confess that at first I regarded it with great doubt and, hoping to disprove it, I started to consider the simplest special cases. But in these cases it proved to be correct if, when $x = y = 0$, we consider $x/y = 1$.

I carried out some verification, far from complete, for the general case. Your equality $E\,xy^n = E\,y^{n+1}$ troubles me. For what values of n have you proved it? If one introduces negative values of $n < -1$, then the right hand side becomes ∞.

If, on the other hand, you assume n is an integer and positive, then it is not clear how to prove the equality $E\,\dfrac{x}{y} = 1$ from this.

I started to verify this very equality, trying to show that

$$E\,\frac{x}{y} = (p + q)^{\sigma s}.$$

The first three terms are obtained fairly easily:

$$p^{\sigma s},\ \sigma s \cdot p^{\sigma s-1} \cdot q,\ \frac{\sigma s(\sigma s - 1)}{1 \cdot 2} \cdot p^{\sigma s-2} \cdot q^2,$$

which also led me to believe that your interesting proposition is true. If you will send me a rigorous proof of it I shall willingly report on it to the

Academy and it can be published in the *Proceedings of the Academy of Sciences.*

Today I received from Bortkiewicz an article containing his retort to Lucy Whitaker. Although I regard the "law of small numbers" with great doubt and even articulated such an opinion in a notice which should have appeared in the *Actuarial Review,* here Bortkiewicz's opposition is apparently not without foundation. Pearson's school seems to me highly dubious. The examples with which Bortkiewicz established his law are unconvincing and can testify only to his cleverness. But I admit that his research on the coefficient of dispersion, while also not fully accurate, is significant. The formula relating to variable dispersion does not cover all cases but nevertheless it can apply to many cases. Pearson's formulas seem useless to me and in the highest degree complex. At present I am reviewing my observations on vowels and I am observing something like the law of small numbers. The coefficients of dispersion for separate letters turn out to be nearer to one than for the whole aggregate of vowels. But for the rarely encountered letter ю this coefficient of dispersion turns out to be significantly larger than one. My eyes, which are refusing to look, impede the calculation. If it will not be too much trouble for you, please send the issue of the publication of the Central Statistical Committee that I mentioned. I want to compare it with that given in the *Medical Journal.*

Sincerely yours,

A. MARKOV

No. 78

(Postcard from Markov to Chuprov)

23 March 1916

Highly respected Alexander Alexandrovich:

The meeting of the Academy of Sciences will be on Wednesday of next week, 30 March. According to the typographical table given in N. A. Morozov's article "Linguistic Spectrum," in German the letter e is encountered particularly often and after that n and this explains why the coefficient of dispersion is less than one for e. For Q I have a coefficient of dispersion less than one but very close to one.

Yours,

A. MARKOV

Is my remark that you have to take the square root of Q^2 not correct?

No. 79

(Postcard from Markov to Chuprov)

23 March 1916

Highly respected Alexander Alexandrovich:

I have succeeded in finding a fairly simple proof of your proposition which has greatly excited my curiosity:

$$E \frac{x}{y} = 1 = (p + q)^{s\sigma}$$

(exactly in this form).

I am not communicating it to you for the present, wishing to know yours beforehand.

Sincerely,

A. MARKOV

No. 80

(Letter from Chuprov to Markov)

24 March 1916

Highly esteemed Andrei Andreevich:

The path you have selected is of course the most direct and natural one. You are obviously proceeding thus. If all t_i are equal to one then $x = 0$ and $y = 0$, and $\frac{x}{y} = 1$; the probability of this is $p^{s\sigma}$. If all t_i are equal to zero then x and y are equal to zero, $\frac{x}{y} = 1$; the probability is equal to $q^{s\sigma}$. If one of the t_i equals one and the remaining equal 0 then

$$t = \frac{1}{s\sigma}, y = \frac{1}{s^2\sigma},$$

the sum of the corresponding x multiplied by the probabilities is equal to

$$C_{s\sigma}^1 \cdot p \cdot q^{s\sigma-1} \cdot \frac{1}{s^2\sigma};$$

hence, the corresponding group of the summands in $E\dfrac{x}{y}$ equals

$$C^1_{s\sigma} \cdot p \cdot q^{s\sigma-1}.$$

For $t = 2/s\sigma$ the corresponding group of summands turns out similarly to be equal to

$$p^2 q^{s\sigma-2} \cdot \frac{s\sigma - 2}{s} \div \frac{2(s\sigma - 2)}{s^2\sigma(s\sigma - 1)},$$

that is,

$$C^2_{s\sigma} \cdot p^2 \cdot q^{s\sigma-2}, \text{ etc.}$$

It goes without saying that this is also a genuine proof for the case $E\,Q^2$. But I took a roundabout path because I approached it from a more general problem.

Statisticians [assume—Editor] that $E\dfrac{x}{y} = \dfrac{E\,x}{E\,y}$, not only in the case of Q^2 but also in others (for example, with the correlation coefficient if we shape the Pearsonian arguments in a stricter form—the problem which I have been studying lately). I have also been trying to establish the conditions under which a similar relation can take place, or in any case to find ways of estimating the difference between $E\dfrac{x}{y}$ and $\dfrac{E\,x}{E\,y}$. I proceeded in just this way. Based on the identity

$$\frac{1}{y} = \frac{1}{b} - \frac{y - b}{by}$$

(I write $E\,x = z$, $E\,y = b$), one easily obtains the equality

$$E\frac{x}{y} = \frac{a}{b} - \frac{1}{b} E\frac{x(y - b)}{y};$$

$$E\frac{x}{y} = \frac{a}{b} - \frac{E\,x(y - b)}{b^2} + \frac{1}{b^2} E\frac{x(y - b)^2}{y} \text{ etc.}$$

and from this, in the case where x/y does not change sign (for example, is always positive as in our case), one of the bounds is in the form:

$$E\frac{x}{y} > \frac{a}{b} - \frac{E\,x(y - b)}{b^2} \quad \text{or} \quad E\frac{x}{y} > \frac{a}{b} - \frac{E\,xy}{b^2}; \qquad (1)$$

$$E\frac{x}{y} > \frac{a}{b} - \frac{E\,x(y - b)}{b^2} + \frac{E\,x(y - b)^2}{b^3} - \frac{E\,x(y - b)^3}{b^4} \dots \text{ etc.}$$

From (1) it follows, by the way, that for $\dfrac{x}{y} > 0$ and with the availability of a reciprocal relation between x and y (i.e. with $E\,xy < 0$) $E\,\dfrac{x}{y}$ is always larger than $\dfrac{E\,x}{E\,y}$. The opposite inequality could be easily obtained if it were possible to prove that

$$E\,\frac{x(y-b)^{2n+1}}{y} > 0$$

also. But you will not prove this in the majority of cases (indeed it will not be true). Then it is possible to try another path. It is possible to replace x/y by a variable chosen so that all of its values are less than r. If for this r is a fixed number, then it comes directly outside the expectation sign. In the case of Q^2, for r one can take $\dfrac{s\sigma - 1}{\sigma - 1}$ or $\dfrac{s\sigma - 1}{\sigma - 1}\left(1 - \dfrac{4\Sigma t_i(1 - t_i)}{\sigma}\right)$. True, this gives extremely wide bounds but nevertheless permits us to come to certain conclusions. I began with this. And then, trying to apply the inequality to the case of Q^2, I happened upon this:

$$E\,xy = E\,y^2;\; E\,xy^2 = E\,y^3;\; E\,xy^3 = E\,y^4.$$

Hence the conjecture occurred to me, will it not be true that $E\,xy^k = E\,y^{k+1}$ for every positive integer k (if I wrote in my letter to you "any k", this was a slip: I always have in mind only positive values) and is this not the root of the fact that, in the given case $E\,\dfrac{x}{y} = \dfrac{E\,x}{E\,y}$.

In my proof of the proposition $E\,xy^k = E\,y^{k+1}$ as well as in the deduction of the formula for the mathematical expectation of any positive integral power of the difference between probability and frequency, I use the properties of the coefficients in the expansion of $x(x-1)(x-2)\ldots(x-n+1)$ in powers of x and in the expansion of x^n by factorials. These properties were partly known long ago; three chapters of *Analyse der Refract* by Kramp are devoted to the study of them; works on Bernoulli numbers touch on them incidentally; a series of Italian works from the school of Cesaro and Cantelli deal with them. But I have to deduce anew a large part of those more complex properties of these coefficients that I need directly. As a result I obtain the following situation: those demonstrations that properly interest me are not too complicated if we assume as given the properties of the aforementioned coefficients that I need for the proof. But since I have to establish these properties myself for the purposes of my work, the complete proofs turn out to be extraordinarily lengthy, and thus the appendix devoted to the study of the coefficients will be longer than the text itself.

The proof that $E\,xy^k = E\,y^{k+1}$ goes like this. I introduce the notation:

$$x(x+1)(x+2)\ldots(x+n-1) = \sum_{k=0}^{n-1} \beta_{n,k} x^{n-k};$$

$$x^n = \sum_{k=1}^{n} \alpha_{n,k} x(x-1)\ldots(x-k+1).$$

With this notation the expectation of a power of the frequency can be represented in the form:

$$E\,t_{so}^r = \sum_{k=1}^{r} p^k \cdot \sum_{i=1}^{k} \frac{(-1)^{k-i}}{s^{r-i}\cdot\sigma^{r-i}} \cdot \alpha_{r,k}\beta_{k,k-i}.$$

Hence:

$$E\,y^{k+1} = \frac{\sigma^{k+1}}{(s\sigma-1)^{k+1}} \cdot E\,t^{k+1} \cdot (1-t)^{k+1}$$

$$= \frac{\sigma^{k+1}}{(s\sigma-1)^{k+1}} \cdot \sum_{h=1}^{2k+2} p^h \cdot \sum_{j=0}^{2k+1} \frac{(-1)^{k+h-j}}{(s\sigma)^{2k+1-j}} \cdot \sum_{l=0}^{k+1} C_{k+1}^{l}$$

$$\cdot\, \alpha_{k+l+1,h} \cdot \beta_{h,k+h}.$$

The summations for $h < k+1$ and $h > k+1$ should properly be carried out not with the same limits but rather with different ones; however, for simplicity in writing one can, without error, write it thus because outside the necessary bounds the variable after the last summation sign can be treated as zero. $E\,xy^k$ also is reduced to this form after certain transformations. But here the calculations are more cumbersome and it is more convenient to present them on separate pages. I shall send them to you tomorrow. It is necessary to rewrite them more legibly which is not so easy for me with my handwriting. As for the transition from $E\,xy^k = E\,y^{k+1}$ to $E\,\dfrac{x}{y} = 1$, first of all it is not difficult to obtain:

$$E\,\frac{x(y-b)^n}{y} = E\,(y-b)^n + (-1)^n \cdot b^2 \cdot E\,\frac{x-y}{y}$$

and from there $E\left\{\left(\dfrac{x-y}{y}\right)\left(\dfrac{y-b}{b}\right)^n\right\}$

$$= (-1)^n \cdot E\,\frac{x-y}{y}.$$

I continue the argument in the following way. The random variables x and y depend on a series of variables (in the case of Q^2, on the t_i) which can take certain values with some probabilities or other. Let the total number of different conceivable systems of values of the variables be equal to μ (in the case of Q^2, $\mu = (s+1)\sigma$); let us denote the probabilities of the separate systems by $p_1, p_2, \ldots, p_\mu$; let us denote the corresponding values of $\dfrac{x-y}{y}$ by $f_1, f_2, \ldots, f_\mu$, and the values of $\dfrac{y-b}{b}$ by $\varphi_1, \varphi_2, \ldots, \varphi_\mu$. We have

$$E\,\frac{x-y}{y} = \sum_{i=1}^{\mu} f_i \cdot p_i; \quad E\left(\frac{y-b}{b}\right)^n = \sum_{i=1}^{n} \varphi_i^n p_i;$$

$$E\left[\left(\frac{x-y}{y}\right)\left(\frac{y-b}{b}\right)^n\right] = \sum_{i=1}^{n} f_i \cdot \varphi_i^n \cdot p_i$$

$$= (-1)^n \cdot \sum_{i=1}^{\mu} f_i \cdot p_i.$$

If some of the variables φ are equal to zero they drop out of the sum. If some of them are equal to each other one can combine them in the sums.

After this let there remain ν variables not equal among each other and not equal to zero. Then the relations written above are reduced to

$$\sum_{j=1}^{\nu} a_j \cdot \varphi_j^n = \sum \left(\frac{y-b}{b}\right)^n; \quad \sum_{j=1}^{\nu} b_j \cdot \varphi_j^n = (-1)^n \cdot \sum_{i=1}^{\mu} f_i p_i.$$

But in order that $\displaystyle\sum_{j=1}^{\nu} b_j \cdot \varphi_j^n = (-1)^n \cdot \sum_{i=1}^{\mu} f_i p_i$ for any whole positive n it is necessary that:

1. at least one of the determinants formed by $\gamma + 1$ rows of the matrix

$$\begin{pmatrix} \varphi_1 & \varphi_2 & \cdots & \varphi_\nu \\ \varphi_1^2 & \varphi_2^2 & \cdots & \varphi_\nu^2 \\ \cdot & \cdot & & \cdot \\ \cdot & \cdot & & \cdot \\ \cdot & \cdot & & \cdot \\ \varphi_1^n & \varphi_2^n & \cdots & \varphi_\nu^n \end{pmatrix}$$

be different from zero;

2. no determinant formed by $\nu + 1$ rows of the matrix

$$
\begin{pmatrix}
-\Sigma \cdot p_i & \varphi_1 & \varphi_2 & \cdots & \sigma_\nu \\
\Sigma f_i \cdot p_i & \varphi_1^2 & \varphi_2^2 & & \varphi_\nu^2 \\
 & \cdot & \cdot & & \cdot \\
 & \cdot & \cdot & \cdots & \cdot \\
 & \cdot & \cdot & \cdots & \cdot
\end{pmatrix}
$$

be different from zero. The first condition together with $\Sigma a_{j\nu}\varphi_j^n =$ $\displaystyle\sum \left(\frac{y-b}{b}\right)^n$ is not of particular interest. The second reduces to

$\Sigma f_i p_i = 0$. But if $\Sigma f_i p_i = E\,\dfrac{x-y}{y} = 0$ then $E\,\dfrac{x}{y} = 1$. The condition

$E\,xy^k = E\,y^{k+1}$ is seen in that way to be sufficient for $E\,\dfrac{x}{y} = 1$.

Whether or not it is necessary I cannot say for the present. Probably not. It is likely that there will also be other forms of dependence

between x and y for which $E\,\dfrac{x}{y} = 1$. I have not settled this yet.

In general outline this is the course of my reasoning. Of course I would be extremely flattered if you would find this problem I am considering of sufficient interest that you would present it to the Academy. As for my solutions—besides their (I fear) somewhat primitive character—what is to be done with that part of the argument where I depend on the properties of the coefficients α and β? Perhaps it is more correct to wait a little with these and then to publish everything together as a whole, including the necessary work on these.

I too received Bortkiewicz's work. There is no doubt that it is interesting and in its general parts, in its overall objections to Pearson and his school it is in many respects valid. But in the parts devoted directly to the defense of the law of small numbers against Whitaker's objection, it is a complete failure in my opinion. Bortkiewicz replies here, not to objections actually made, but rather produces quite another argument. Also the general estimate of Whitaker's method is given totally incorrectly. Take for example the reductio ad absurdum on page 230. Here Bortkiewicz's jump in thought is for me frankly incomprehensible. For any p and q we have, in expectation

$$
p = \frac{\sigma^2}{m},\; q = \frac{m-\sigma^2}{m},
$$

$$
n = \frac{m^2}{m-\sigma^2} \quad \text{and} \quad Q^2 = \frac{\sigma^2}{npq} = 1 \quad \text{or} \quad Q^2 p = p.
$$

The relation $Q^2 p = p$ remains valid also for p near one. Bortkiewicz, passing to the case of small q, takes p equal to one on the left side but on the right, $p < 1$. Naturally one gets an absurdity but neither Whitaker nor her method is guilty here. The arguments on page 238 are also incorrect, as if Whitaker's method shifts the problem. Whitaker's method leads to exactly the same correspondence of initial hypotheses with empirical data as does the calculation of Q^2, but also has the advantage that we can by consideration of higher moments extend the verification. But of course it is necessary in using them to take account of the probable errors in the determination of variables p, q, n. And Whitaker does this quite consistently all the time. And if Bortkiewicz would fit corresponding squared errors to the results of the examples he cites on pages 253–256 he would then also obtain both for p and for n exactly those percentage deviations from the actual variables as for Q^2 (and as it should be, with satisfaction of all of the conditions of normal dispersion). It is this way too with the inference on page 256, as if these examples must be disconcerting, in contrasting these methods with Lexis's theory of dispersion, for which there is no basis. The methods used by Whitaker do not give rise to any divergence from that which is reasonable in Lexisian theory. In essence this is exactly the same, only the mathematical form is different.

The reproaches on pages 235–236 are also directed to the wrong quarter. Bortkiewicz's calculations are interesting in themselves but they prove only that, with small N, examples cannot be decisive, which Whitaker stresses at all times on all those pages cited by Bortkiewicz. In the treatment of statistical material it is better to compute Q^2 than Q. The habit of computing Q was engaged in following the example of Lexis who simply hadn't thought it through completely in his time.

Sincerely yours,

A. Chuprov

No. 81

(Postcard from Markov to Chuprov)

24 March 1916

Highly respected Alexander Alexandrovich:

Your proposition can also be proved in a more general form for unequal sequences:

$$E \frac{\sum s_i \cdot (t_i - t)^2 \frac{1}{\sigma - 1}}{\frac{s}{s - 1} \cdot t \cdot (1 - t)} = 1 = (p + q)^s.$$

{ a unique fractional expression whose mathematical expectation is very simply found.

where $s = s_1 + s_2 + \cdots + s_\sigma$.

The proof I have found is fairly simple. (The general case is not more complicated than the particular).

Sincerely yours,

A. MARKOV

No. 82

(Letter from Markov to Chuprov)

25 March 1916

Highly respected Alexander Alexandrovich:

Thank you for the tables you sent. In my opinion these tables are a shaky foundation for compiling height curves, not because a difference of one vershok is too great but because at first it was 1/2 vershok and later a whole vershok. If the numbers had been given for 2 ver., 3 ver., 4 ver., ..., I would have understood them thus:

from 3/2 to 5/2 from 5/2, to 7/2 etc.

and would have tried to fit truncated Laplace curves, for here all of the people whose height is less than the norm are discarded. But with data $2\frac{1}{2}$, 3, 4, 5, 6, ..., apparently it is to be assumed that in the first interval there was good accuracy and that's why such intervals were taken.

$2\frac{1}{4} - 2\frac{3}{4}$, $2\frac{3}{4} - 3\frac{1}{2}$, $3\frac{1}{2} - 4\frac{1}{2}$, etc. so that the first interval constitutes 1/2, the second 3/4, and so on.

To me this combination is questionable.

Then, a comparison with the *Army Medical Journal* where data are given for the year 1875 shows disagreement of the numbers. In the journal all of the measurements are given in fourths of vershoks but they refer not only to those admitted into service but to all those examined. I believe that in the *Statisticheskiĭ Vremennik* we have not encountered the original numbers, just as in the *Army Statistical Review* of the Petersburg Military District. The latter was compiled from the numbers in the *Army Medical Journal* and gave numbers in the form:

$$< 34 \text{ ver.}, 34–35, 35–36, \ldots$$

It could be compared with the *Army Medical Journal* to find out how the numbers in the *Army Statistical Review* were obtained. It is impossible to do anything similar for the *Statisticheskiĭ Vremennik* and it remains only to acknowledge some kind of duplicity in our official statistics.

Has no one attempted to construct tables of the height of recruits from such data? It seems questionable to me. In my personal opinion another use could be made of these tables, computing them correctly.

Take a definite height, for example 2 ar., 5 ver., and from the tables determine the probability of such a height in different years.

How does this probability change from year to year? Right now I see only that the height of recruits in Russia can hardly serve as a support for the regular applicability of Laplace's formulas. These heights were established by questionable data. Someone has altered them.[1]

Sincerely yours,

A. MARKOV

---

[1]The basic data were perhaps also poor for in them there are strange fluctuations of the numbers, but no one had the right to change them and I am sure it was done.

No. 83

(Letter from Markov to Chuprov)

26 March 1916

Highly respected Alexander Alexandrovich:

I have convinced myself of the truth of your proposition but your proof frightens me by its complexity. My proof is distinguished by its simplicity and has to do with the general case of σ sequences of $s_1, s_2, \ldots, s_\sigma$ observations.

It is based on a particular system of summation in which those terms where the t's have the same value are combined into one sum.

If x with values:

$$x_1, x_2, \ldots, x_i, \ldots, x_\sigma$$

denotes the number of occurrences of an event, then the numbers t_i will be

$$\frac{x_1}{s_1}, \frac{x_2}{s_2}, \ldots, \frac{x_i}{s_i}, \ldots, \frac{x_\sigma}{s_\sigma}$$

and

$$t = \frac{x_1 + x_2 + \cdots + x_\sigma}{s_1 + s_2 + \cdots + s_\sigma} = \frac{\ell}{N}.$$

The expression for Q^2 becomes:

$$Q^2 = \Sigma \, s_i \frac{\left(\dfrac{x_i}{s_i} - \dfrac{\ell}{N}\right)^2 \cdot \dfrac{1}{\sigma - 1}}{\dfrac{\ell}{N-1} \cdot \left(1 - \dfrac{\ell}{N}\right)}$$

$$= \frac{N(N-1)}{\ell(N-1)(\sigma-1)}\left[\Sigma \frac{x_i^2}{s_i} - \frac{\ell^2}{N}\right]$$

and

$$E \, Q^2 = P_{x_1, x_2, \ldots, x_\sigma} Q^2,$$

where $P_{x_1, x_2, \ldots, x_\sigma}$ is the probability of the system of numbers $x_1, x_2, \ldots, x_\sigma$ and is expressed by a well-known product.

I combine into one group those terms for which $x_1 + x_2 + \cdots + x_\sigma = \ell$ maintains one value, excluding beforehand the cases $\ell = 0$ and $\ell = N$ whose probabilities are expressed by the powers q^N and p^N.

For the computation of each sum $\Sigma \, P_{x_1, x_2, \ldots, x_\sigma} \cdot Q^2$ where $x_1 + x_2 + \cdots + x_\sigma = \ell$ has a definite value, I replace the product

$$(p\xi_1 + q)^{s_1} \cdot (p\xi_2 + q)^{s_2} \ldots (p\xi_\sigma + q)^{s_\sigma},$$

whose coefficient for

$$\xi_1^{x_1}, \xi_2^{x_2}, \ldots, \xi_\sigma^{x_\sigma}$$

is equal to $P_{x_1,x_2,\ldots,x_\sigma}$, by $(p\xi_1 \cdot Q + q)^{s_1} \cdot (p\xi_2 \cdot Q + q)^{s_2} \ldots (p\xi_\sigma \cdot Q + q)^{s_\sigma}$. This last product is expressed by the sum:

$$\Sigma\, P_{x_1,x_2,\ldots,x_\sigma} \cdot \xi_1^{x_1} \cdot \xi_2^{x_2} \ldots \xi_\sigma^{x_\sigma} \cdot Q^{x_1+x_2+\cdots+x_\sigma},$$

so that to combine all the $P_{x_1,x_2,\ldots,x_\sigma}$ where $x_1 + x_2 + \cdots + x_\sigma = \ell$ into one group, it is necessary to take the terms in the product with Q^ℓ. Thus $\Sigma\, P_{x_1,x_2,\ldots,x_\sigma}$, with $x_1 + x_2 + \cdots + x_\sigma = \ell$, turns out to be equal to the coefficient of Q^ℓ in the expansion of $(pQ + q)^N$ in powers of Q:

$$\sum_{x_1+x_2+\cdots+x_\sigma=\ell} P_{x_1,x_2,\ldots,x_\sigma} = \frac{N(N-1)\cdots(N-\ell+1)}{1\cdot 2 \ldots \ell}\, p^\ell \cdot q^{N-\ell}.$$

Thereupon, in order to find $\displaystyle\sum_{x_1,x_2,\ldots,x_\sigma} P\, x_i^2$ under the same conditions it is

necessary to set $\eta = 0$ in the expression $(pQ + q)^{N-s_i} \dfrac{d^2(p\ell^\eta \cdot Q + q)^{s_i}}{dy^2}$
and find the coefficient of Q^ℓ in the expression $s_i \cdot p \cdot Q(pQ + q)^{N-1} + s_i(s_i - 1) \cdot p^2 \cdot Q^2 \cdot (pQ + q)^{N-2}$ which turns out to be equal to

$$s_i \frac{(N-2)\cdots(N-\ell+1)}{1\cdot 2 \ldots (\ell-1)} \{N - 1 + (s_i - 1)(\ell - 1)\}\, p^\ell \cdot q^{N-\ell}.$$

Consequently

$$\sum_{x_1+x_2+\cdots+x_\sigma=\ell} P_{x_1,x_2,\ldots,x_\sigma} \cdot \left(\frac{x_1^2}{s_1} + \frac{x_2^2}{s_2} + \cdots + \frac{x_\sigma^2}{s_\sigma} \right)$$

reduces to

$$\frac{(N-2)\cdots(N-\ell+1)}{1\cdot 2 \ldots (\ell-1)} \{\sigma(N-1) + (N-\sigma)(\ell-1)\} \cdot A,$$

$$A = p^\ell \cdot q^{N-\ell}.$$

Computing from here:

$$\frac{\ell^2}{N} \sum_{x_1+x_2+\cdots+x_\sigma=\ell} P_{x_1,x_2,\ldots,x_\sigma} = \frac{(N-1)(N-2)\cdots(N-\ell+1)\ell}{1\cdot 2 \ldots (\ell-1)} \cdot A,$$

we find:

$$\sum P_{x_1,x_2,\ldots,x_\sigma} \cdot \left(\frac{x_1^2}{s_1} + \frac{x_2^2}{s_2} + \cdots + \frac{x_\sigma^2}{s_\sigma} - \frac{\ell^2}{N} \right)$$

$$
= \frac{(N - 2)(N - 3) \cdots (N - \ell + 1)}{1 \cdot 2 \ldots (\ell - 1)}
$$
$$
\times \{(\sigma - \ell)(N - 1) + (N - \sigma)(\ell - 1)\} \cdot A
$$
$$
= \frac{(\sigma - 1)(N - 2)(N - 3) \cdots (N - \ell + 1)(N - \ell)}{1 \cdot 2 \cdots (\ell - 1)} \cdot p^\ell \, q^{N-\ell}
$$

and then

$$
\sum_{x_1 + x_2 + \cdots + x_\sigma = \ell} P_{x_1, x_2, \ldots, x_\sigma} \cdot Q^2 = \frac{N(N - 1) \cdots (N - \ell + 1)}{1 \cdot 2 \ldots (\ell - 1)\ell} \, p^\ell \cdot q^{N-\ell}.
$$

This result relates to all values of $x_1 + x_2 + \cdots + x_\sigma = \ell$ except 0 and N. Assuming $Q^2 = 1$ in these exceptional cases, we extend the formula to them also. It remains to give ℓ all the values $0, 1, 2, \ldots, N$ and to carry out the summation which yields

$$
\sum P_{x_1, x_2, \ldots, x_\sigma} \cdot Q^2 = \sum \frac{N(N - 1) \cdots (N - \ell + 1)}{1 \cdot 2 \ldots (\ell - 1)\ell} \, p^\ell \cdot q^{N-\ell}
$$
$$
= (p + q)^N = 1.
$$

The method indicated also gives your general proposition

$$
E \, xy^n = E \, y^{n+1}
$$

for all values of n, not only integral but also fractional.

Sincerely yours,

A. MARKOV

No. 84

(Letter from Markov to Chuprov)

29 March 1916

Highly respected Alexander Alexandrovich:

Today I received your calculations; I will confess that they scare me with their complexity. Possibly they have much in common with that which I sent you the other day.

The result at which I arrive can be expressed thus: for each given value of y the equality

$$\Sigma \, P \cdot x = y \, \Sigma \, P,$$

holds where P denotes the probability of a system of numbers x and y.

Hence, on the one hand all of the equalities

$$\Sigma\Sigma \, P \, xy^n = \Sigma\Sigma \, y^{n+1} \cdot P$$

for any positive number, even if it is not an integer, follow; and on the other hand—if for $y = 0$ the number x also equals zero and *we set* the ratio $x/y = 1$, then dividing by y we can write

$$\sum P\frac{x}{y} = \sum P \text{ and } \sum\sum P\frac{x}{y} = \sum\sum P = 1.$$

It would be a great pity if my letter containing this result were lost. Its basic idea, which was carried out for sequences of different numbers of observations, consisted of replacing the product

$$(pQ_1 + q)^{s_1} \cdot (pQ_2 + q)^{s_2} \ldots (pQ_\sigma + q)^{s_\sigma}$$

by the product

$$(pQ_1 \cdot Q + q)^{s_1} \cdot (pQ_2 \cdot Q + q)^{s_2} \cdots (pQ_\sigma \cdot Q + q)^{s_\sigma}.$$

By the way, once the formula

$$\Sigma\Sigma \, P \, xy^n = \Sigma\Sigma \, P \, y^{n+1}$$

is proved for all positive integer values it is easy to prove the formula

$$\Sigma \, P \, x = y \, \Sigma \, P,$$

if the number of different possible values of y is finite, as in fact it really is. Let $y_1, y_2, \ldots, y_m$ be all of the different values of y. Furthermore, let M_i denote the difference

$$\Sigma Px - y \, \Sigma \, P \quad \text{for} \quad y = y_i.$$

Then we can write a series of equalities:

$$M_1 + M_2 + \cdots + M_m = 0,$$

$$M_1 y_1 + M_2 y_2 + \cdots + M_m y_m = 0,$$

$$\cdots\cdots\cdots\cdots\cdots\cdots\cdots\cdots\cdots$$

$$M_1 y_1^{m-1} + M_2 y_2^{m-1} + \cdots + M_m y_m^{m-1} = 0,$$

and since the determinant

$$\begin{vmatrix} 1 & 1 & \cdots & 1 \\ y_1 & y_2 & \cdots & y_m \\ \cdot & \cdot & & \cdot \\ \cdot & \cdot & & \cdot \\ \cdot & \cdot & & \cdot \\ y_1^{m-1} & y_2^{m-1} & \cdots & y_m^{m-1} \end{vmatrix}$$

is different from zero we must have $M_1 = M_2 = \cdots = M_m = 0$.

Your proposition interests me greatly but the proofs are too complicated and in part not completely convincing. After it is proved that $E \dfrac{x}{y} = 1$ the question arises about $E \left(\dfrac{x}{y} \right)^2$ and $E \left(\dfrac{x}{y} - 1 \right)^2$. Unfortunately the general expression for $E \left(\dfrac{x}{y} \right)^2$ turns out to be complicated.

The method of successive summations can also be applied here but the second summation does not lead to a simple general formula, at least it seems so to me.

I turn to the connection between the particular problem and the general question about the existence of the equality

$$E \frac{x}{y} = \frac{E \, x}{E \, y}$$

and its corollary

$$\frac{x}{y} \approx \frac{E \, x}{E \, y}.$$

Of course it is impossible to write similar equalities in the general case but in application to Lexis's case one can justify the second equality.

If, denoting $E\,x$ by a and $E\,y$ by b, we prove that $E\left(\dfrac{x}{a}-1\right)^2 < \alpha$ and $E\left(\dfrac{y}{b}-1\right)^2 < \beta$, then, with probability greater than $1-\dfrac{2}{t^2}$, we can assert that, simultaneously

$$1 - t\sqrt{\alpha} < \frac{x}{a} < 1 + t\sqrt{\alpha}$$

$$1 - t\sqrt{\beta} < \frac{y}{b} < 1 + t\sqrt{\beta}$$

and consequently the ratio x/y has the bounds

$$\frac{a}{b}\cdot\frac{1 - t\sqrt{\alpha}}{1 + t\sqrt{\beta}} \quad \text{and} \quad \frac{a}{b}\cdot\frac{1 + t\sqrt{\alpha}}{1 - t\sqrt{\beta}}.$$

In those cases where α and β are very small numbers, as in Lexis's problem, with large s and σ, it is then possible, with high probability, to conclude that x/y is close to a/b.

One must assume that for Lexis's case it is possible to establish some kind of not too complicated inequality

$$E\,(Q^2 - 1)^2 < \epsilon$$

from which it will follow that with high probability Q^2 will be close to 1. But the same, although with smaller probability, can be established by considering separately the numerator and denominator of x/y which is much simpler. I have not seen L. Whitaker's article and moreover it is written in a language unknown to me. However, the examples pertaining to Meissner's experiments which Bortkiewicz indicated at the end of his article impress me. Bortkiewicz points out that it is possible to obtain absurd results by the path of Whitaker and Pearson (who taught and inspired her) in other examples as well.

Sincerely yours,

A. Markov

I consider Bortkiewicz's examples meaningless because the number of series is small and because he chose material that was pleasing to him. Also, for small numbers one cannot expect large coefficients of dispersion.

I observe that Bortkiewicz assumes that the probabilities are variable so that it is wrong to apply formulas for constant probability to his examples.

Yours,

A. MARKOV

I have 2000 sequences of 100 observations and 400 sequences of 500 observations.

No. 85

(Letter from Markov to Chuprov)

31 March 1916

Highly respected Alexander Alexandrovich:

Of course I fully acknowledge the value of your computations which led to a remarkable result. I just think that if you want to study $E\,(Q^2 - 1)^2$ it is necessary to think about simplifying the calculation which is achieved by the method I indicated. An exact expression for $E\,(Q^2 - 1)^2$ is complicated since only the first summation is fairly simple. But in any case it remains to carry out a summation of a certain type:

$$\Sigma\, H_\ell \cdot p^\ell\, q^{\sigma-\ell},$$

where H_ℓ is a certain function of ℓ. And I believe that it is possible to find a more or less adequate inequality giving an upper bound for H_ℓ and at the same time simplifying the sum. As for L. Whitaker's criticism, as far as I can judge from what Bortkiewicz said and from what is in your letters, I consider it wrong. In all cases where q is very small Whitaker's method must give a questionable result, which is obvious also from the expression

$$\frac{np}{\sqrt{\pi}\cdot q}\,\sqrt{2\left(1 - \frac{1}{n}\right)}$$

you mentioned which, by the way, is clear by itself. Indeed, Bortkiewicz has just such cases in mind too. In his questions only the product nq, whose factors are inseparable, plays a role. His number q is perhaps less than 0.0001. Under such conditions the formula

$$p = \frac{\sigma^2}{n}$$

can easily give a number greater than one, i.e., nonsense. It can yield the correct result only under the condition that the coefficient of dispersion is exactly equal to its mathematical expectation, i.e., one. If this coefficient deviates from one on the positive side by only 0.001 it will be greater than one. If Whitaker does not arrive at nonsense in the other examples, then it is only because q is comparatively large.

Bortkiewicz just cites examples where q although not as small as in his questions is small enough that the weakness of Whitaker's method is revealed. Thus, where the coefficient of dispersion is less than one $\left(\dfrac{A}{B} \right)$, the size of q comes out significantly larger than it should but n, on the contrary, less than it should. In example C where $Q^2 > 1$, L. Whitaker's method gives too small a value for q but gives $n = 2629$ instead of 180. Finally, in example D, the same sort of nonsense is obtained as in Bortkiewicz's cases. The incorrectness of Whitaker's method is clear to me. She should have thought about her expression

$$\frac{np}{\sqrt{\pi \cdot q}} \cdot \sqrt{2 \left(1 - \frac{1}{n} \right)}$$

where q is an unknown but very small number. Then, one must remember that Bortkiewicz relates his law to variable probabilities since Bortkiewicz's formula is concerned with variable probability. Of course, he himself partly forgets this circumstance, probably considering q (or nq) as a variable that changes little. However, it is impossible to disregard completely this circumstance with which Bortkiewicz's law of small numbers is closely connected but, it seems to me, this is done by L. W. with whose work I am acquainted only through Bortkiewicz's objections and your letters.

Sincerely yours,

A. MARKOV

Isn't one of your students whom you have mentioned coming to my place? I am always at home from 6:00–7:00 o'clock.

No. 86

(Letter from Markov to Chuprov)

1 April 1916

Highly respected Alexander Alexandrovich:

Chebyshev's inequality is remarkably useful for theoretical purposes but generally speaking it is useless for practical applications as is mentioned, among other things, on page 73 of my book. However, it is impossible to establish a different inequality on the basis of the mathematical expectation of the square unless a definite probability law is used. Consideration of mathematical expectations of higher powers brings in more complexity but yields a little. If you consider the well-known formula of Gauss, the fraction $\frac{1}{t^2}$ is replaced by the integral

$$\frac{2}{\sqrt{2\pi}} \int_{t/\sqrt{2}} e^{-t^2}\, dt,$$

which is already less than 0.0001 for $t = 3\sqrt{2}$.

Sincerely yours,

A. MARKOV

One must assume that the value of $E\left(\dfrac{x}{y} - 1\right)^2$ will also contain fractions such as $\dfrac{2}{\sigma - 1}$ and $\dfrac{1}{s\sigma nq}$. Therefore, for small values of σ, even with the assumption of Laplace's probability law, it is impossible to expect that x/y will be close to one with high probability.

No. 87

(Letter from Chuprov to Markov)

27 September 1916

Highly esteemed Andrei Andreevich:

My article "Mathematical Foundations of the Theory of Statistical Series" is just beginning to be published; two of the first essays and a fairly extensive appendix devoted to the study of the quantities that you call C_m^i and D_m^i in the *Calculus of Finite Differences* will appear in the volume of the *Proceedings of the Polytechnic Institute* which will come out this fall. Of course, as soon as the reprints come I will let you know promptly; your opinion is of the greatest interest to me. As to the fact that $EQ < 1$, it was pointed out by Bortkiewicz in the article "Der Wahrscheinlichkeitstheoretische" (*Oesterreichische Revue*, Wien, 1906) in which he arrives at the formula $EQ = 1 - \dfrac{1}{4\sigma}$. Since then Bortkiewicz himself and other investigators often do not compute Q, confining themselves to the computation of Q^2, but not from those considerations of principle according to which you do not go from EQ^2 to EQ and which I too support (the fact that the precise value of EQ is unknown), but in order to avoid unnecessary work. Right now I do not have this literature at hand but if it interests you I can find it. Bortkiewicz's reasoning which leads to the relation $EQ = 1 - \dfrac{1}{4\sigma}$ can in no way be considered a proof since the result is obtained by arbitrarily dropping from the equation terms of the same order as those that are retained. In my article which is being published in *Proceedings of the Polytechnic Institute* I examine their analysis and indicate the correct method of solution of the problem posed by Bortkiewicz. Observing that $Q^2 \geq 0$ and stipulating that the plus sign be taken for Q^2, we can use all of the inequalities that follow from Liapunov's lemma (or from Stieltjes' conditions). I find by this method that

$$\sqrt{1 - \frac{2}{\sigma - 1}} < E\,Q < 1.$$

Until now I have not succeeded in obtaining more advantageous bounds. The transition from one of the inequalities obtained from your formula on page 715 to the other essentially adds almost nothing, of course. But the formulation of the result $E(Q^2 - 1) < \dfrac{2}{\sigma - 1}$ becomes a little nicer since, instead of the restriction "for $\sigma \geq 5$," it turns out to be possible to confirm the inequality for any σ (in fact it no longer arises for $\sigma = 1$). Only in this do I see the interest in such a change.

In my last formulas relating to EQ^4 the problem of finding C is easily reduced to finding $E\,\dfrac{(x_1 - x_2)^4}{y^2}$ or $E\,\dfrac{(x_1 - x_2)^2(x_1 - x_4)^2}{y^2}$, finding A is reduced to $E\,\dfrac{(x_1 - x_2)^2(x_3 - x_4)^2}{y^2}$, $B = C - 6A$ etc.

One can derive many diverse relations. I have not succeeded in clarifying the sign of B; apparently it depends on the sign of $\mu_4 - 3\mu_2^2$, but I have not succeeded at all in proving this rigorously. But if one could find the sign of B, then for the case of sequences of different length the problem would be quite satisfactorily solved. For $B < 0$ even for sequences of different length:

$$E \, Q^4 < 1 + \frac{2}{r-1} \cdot \frac{s(s-r)}{(s-2)(s-3)}.$$

I am not losing hope of completing this. But there is no time now to work on it. Of the various inequalities I have obtained (for sequences of different length) the following is not without interest:

$$E \, Q^4 > 1 + \frac{2}{r-1} \cdot \frac{r(n-1)(nr-1)}{n^2 r^2} \cdot 16 p^2 \cdot q^2,$$

It gives a lower bound for EQ^4; for $p = 1/2$ we find:

$$E \, Q^4 > 1 + \frac{2}{r-1} \left(1 - \frac{1}{n} \right) \left(1 - \frac{1}{nr} \right).$$

With regard to the derivation of the statement $EQ^2 = 1$, it is of course possible to carry out the reasoning in a straightforward manner also for $k = -1$, not going by the detour of arguing that $E \, \dfrac{x}{y} = 1$, if $E \, xy^k = E \, y^{k+1}$ for $k = 0, 1, \ldots$ I indicate this in one of the notes where I also speak about the fact that my method of deduction is not applicable to $E \, \dfrac{x}{y}$; but in finding EQ^4 a circuitous route cannot be avoided. In view of this I also was inclined to base the proof entirely on calculations. A convention regarding $0/0$ is necessary, of course. As I see it, I made such a convention also in the deduction of EQ^4; here, however, I have not carried it through completely. As of now I have not tackled the computation of EQ^6, EQ^8, etc. The calculations must be very complicated. From the point of view of statistical inquiry these problems are less real than finding EQ^2 and EQ^4. I shall willingly make any changes in my article that you consider necessary. Since I have not kept a rough copy in a convenient form, I would ask you to send me my manuscript for this purpose.

Sincerely yours,

A. CHUPROV

No. 88

(Letter from Chuprov to Markov)

14 October 1916

Highly esteemed Andrei Andreevich:

I have completely omitted the calculations concerning EQ''^2 and almost completely omitted those concerning EQ'^2. Perhaps it would be possible to omit them entirely too, reducing page 19 simply to "Similarly we find: $EQ'^2 = 1$." If you agree, would you be so kind as to cross out the intervening words.

I am thinking of entitling it in French "Sur l'espérance mathématique du coefficient de divergence."

Sincerely yours,

A. CHUPROV

$$\frac{1}{\pi} \int_{-\infty}^{+\infty} \frac{\sin \alpha \xi}{\xi} \cdot e^{-\lambda^2 \xi^2} \, d\xi = \frac{2}{\sqrt{2\pi}} \int_{0}^{\alpha/2\lambda} e^{-t^2} \, dt.$$

No. 89

(Letter from Markov to Chuprov)

28 January 1917

Highly respected Alexander Alexandrovich:

Not having verified your calculations I have focused my attention on some of your results and on their connection with Pearson's formula which you consider applicable here. Unfortunately you have not indicated the main point: how x in Pearson's formula is connected with $Q^2 - 1$. Therefore it is not clear to me whether x can and must attain the value $-a$ as Pearson's calculations assume.

If you have a proof that $-a$ is a sharp bound for x, I ask you to do this. For $r = 2$ it is possible that the result can be considered correct if $x = Q^2 - 1$. For $r = 3$ something strange is obtained, and for other values of r I ask you to clarify this point.

As for your computations of the bounds for the mathematical expectations of powers of Q, you have not determined them beforehand in a different way for the normal case when only the number of observations in each sequence varies. In fact for this case it is possible to take Bort-kiewicz's expression for Q, i.e. to consider the denominator constant.

If r is the number of sequences and n is the number of observations in each sequence then we can set the numbers of occurrences of an event equal to $np + t_1 \cdot \sqrt{2npq}$, $np + t_2 \cdot \sqrt{2npq}, \ldots$ and in the expression for Q^2 disregard the $\sqrt{n}$ as compared to n in the denominator in order to obtain the expression for Q^2 as the quadratic form:

$$2 \frac{(r-1)\Sigma t_i^2 - 2\Sigma t_i t_j}{r(r-1)} = Q^2 \qquad \text{(according to you).}$$

Then the mathematical expectations of different powers are the integrals

$$\frac{1}{(\sqrt{\pi})^r} \int\limits_{-\infty}^{+\infty} \int \cdots \int \left\{ 2 \frac{(r-1)\Sigma t_i^2 - 2\Sigma t_i t_j}{r(r-1)} \right\}^k$$
$$\times - dt_1\, dt_2 \cdots dt_r \cdot e^{-t_1^2 - t_2^2 - \cdots - t_r^2}.$$

For $r = 2$ a very simple result is obtained from which one can also arrive at a special case of the Pearsonian formula. However, in other cases the matter becomes very complicated.

Your results, as far as I have checked them (as of now only for $k = 3$), agree with this formula. The transition from it to the expression for the probability seems difficult to me right now. If, together with n, we are to vary other elements, for example p, then the indicated transition may prove to be quite inappropriate. Under the assumption indicated, all attention is focused on the values of m close to np in the usual sense that $\lim \dfrac{m}{np} = 1$. Under Bortkiewicz's assumption $p = q/n$, it is already impossible to treat $(n - m) \cdot m$ as a constant and moreover, the integration is replaced by a summation. For Q^4 it is not difficult to make the appropriate changes if it is a question, as I always assume, of the repetition of identical trials and the counting of the number of occurrences of an event.

Sincerely yours,

A. MARKOV

The other day I happened to become acquainted with the application of the method of least squares to a field trial. In the journal *Economy*, No. 45–46 for 1916, mathematical formulas are given. They are utterly absurd but many people believe them.

No. 90

(Postcard from Markov to Chuprov)

28 January 1917

Highly respected Alexander Alexandrovich:

If you are interested in the limit of the mathematical expectation of $(Q^2 - 1)^2$ under the condition (in the notation of my note) $p = g/s$ where g is constant as is σ but s increases indefinitely, it is not difficult to express in terms of Poisson's formula. However, it is expressed as the sum of an infinite series

$$\frac{2}{\sigma - 1} \cdot e^{-g\sigma} \cdot \left\{ \frac{(g\sigma)^2}{1 \cdot 2^2} + \frac{2(g\sigma)^3}{1 \cdot 2 \cdot 3^2} + \frac{3(g\sigma)^4}{1 \cdot 2 \cdot 3 \cdot 4^2} + \cdots \right\}$$

Yours,

A. MARKOV

It can also be expressed by certain integrals.

No. 91

(Letter from Markov to Chuprov)

29 January 1917

Highly respected Alexander Alexandrovich:

First of all, thank you for pointing out the misprints, some of which I had not yet noticed. It is beyond any doubt that the DeMoivre-Laplace formula should not be applied, even in the limit, to the difference $Q^2 - 1$ since this difference is not less than -1. Consequently, for the objective indicated, there is no need to consider mathematical expectations. For $r = 2$ there is reason for applying the formula $A \int e^{-x} \cdot x^g \, dx$ for the limit of Q^2. Also I consider comparisons with Pearsonian formulas interesting and am very glad that you are doing this. It represents a theoretical example on which the suitability of these formulas can be tested. It is just a pity that these formulas contain too many undetermined constants that one can adjust. Therefore one must consider mathematical expectations of high powers in order to test them.

However, already the question about the number a, which with a minus sign is a bound for the values of x, will perhaps serve to make the applicability of Pearson's formulas questionable. And so I also asked you to indicate to me how Q^2 is linked to x. The easiest case of all is $r = 3$ (after the case $r = 2$ which is not worth talking about) but for it $a = 0$ and Pearson's formula requires some kind of transformation. Thereupon, for me the main question is the equality

$$\lim E \frac{\alpha}{\beta} = \lim \frac{E\alpha}{E\beta},$$

on the basis of which you are apparently making the transition from

$$E \frac{z^3}{q^3} \quad \text{and} \quad E \frac{z^4}{q^4} \quad \text{to} \quad \frac{Ez^3}{Ey^3} \quad \text{and} \quad \frac{Ez^4}{Ey^4}.$$

How and under what conditions do you establish a similar equality? After the transition from the ratio z/y to the separate variables z and y such a simplification is achieved immediately so that it is only necessary to establish clearly the possibility of this transition. Perhaps I missed something but I did not find it for $E \frac{z^2}{y^2}$ either.

After the sealed letter I immediately sent you a postcard on which I gave the limiting value of $E(Q^2 - 1)$ (in your notation) under the condition that $np = g = $ constant and $n = \infty$. I hope that you received it. It is not difficult to arrive at this limit formula. In case you did not receive my postcard I shall repeat it.

$$\lim E \, (Q^2 - 1)^2 = \frac{2e^{-gr}}{r-1} \left\{ \frac{1 \cdot (gr)^2}{1 \cdot 2 \cdot 2} + \frac{2(gr)^3}{1 \cdot 2 \cdot 3 \cdot 3} \right. \\ \left. + \frac{3(gr)^4}{1 \cdot 2 \cdot 3 \cdot 4 \cdot 4} + \cdots \right\}.$$

It can also be expressed as a certain integral. Regarding the Academy honors, I think that they are given out too leniently.

Sincerely yours,

A. MARKOV

No. 92

(Postcard from Markov to Chuprov)

29 January 1917

Highly respected Alexander Alexandrovich:

For $p = m/n$ the limit of $E(Q^2 - 1)^2$ can be represented (in agreement with the earlier expression) as

$$\frac{2}{r-1} \left\{ 1 - e^{-rm} - e^{-rm} \cdot \int_0^{rm} \frac{e^x - 1}{x} \right\} dx.$$

from which it is clear that this limit is always less than $\dfrac{2}{r-1}$ for any rm. From this various inequalities can also be derived.

Yours,

A. MARKOV

No. 93

(Letter from Markov to Chuprov)

30 January 1917

Highly respected Alexander Alexandrovich:

I usually denote your Q^2 by the letter Q since by extracting the square root the matter becomes greatly complicated. The result for $p = g/n$ and $n = \infty$ follows immediately from the formula in my note about the coefficient of dispersion, replacing probabilities by their limiting values according to Poisson's formula. The method I have indicated for finding $\lim E(Q^2)^\ell$ (in your notation) makes it possible to avoid tiresome calculations in which one can easily omit one term or another.

I have no doubt about the correctness of my results, but I confine them to the usual case of probabilities determined from the observations where the denominator of Q^2 is the product of the number of trials and the number of occurrences of the event, if we do not count the constant factor. For this case, first of all, in the notation of your manuscript, the limit of $E\,(Q^2)^\ell$ reduces to the multiple integral

$$\frac{1}{(\sqrt{\pi})^r}\int_{-\infty}^{+\infty}\int\cdots\int \{r\xi-\eta^2\}^\ell \cdot \left\{\frac{2}{r(r-1)}\right\}^\ell \cdot e^{-\xi}\,dt_1\,dt_2\cdots dt_r,$$

where $\xi = t_1^2 + t_2^2 + \cdots + t_r^2$ and $\eta = t_1 + t_2 + \cdots + t_r$. I have not finished the computation of this integral but I am on the right path on which special difficulties should not be encountered. In the first place Newton's binomial theorem

$$(r\,\xi - \eta^2)^\ell = r^\ell \xi^\ell - \ell r^{\ell-1} \cdot \xi^{\ell-1} \cdot \eta^2 + \cdots$$

reduces the computation to integrals of the form

$$\int\int\cdots\int \xi^i \eta^{2j}\,dt_1\,dt_2\cdots dt_r,$$

for the computation of which I first consider

$$\begin{aligned}
\Omega &= \frac{1}{(\sqrt{\pi})^r}\int\int\cdots\int e^{-\alpha\xi+\beta\eta}\,dt_1\,dt_2\cdots dt_r\\
&= \frac{1}{(\sqrt{\pi})^r}\int\int\cdots\int \exp\left[-\alpha\sum\left(t_i - \frac{\beta}{2\alpha}\right)^2 + \frac{r\beta^2}{4\alpha}\right]dt_1\,dt_2\cdots dt_r\\
&= \left(\frac{1}{\sqrt{\pi\alpha}}\right)^r \cdot e^{r\beta^2/4\alpha} \cdot \int_{-\infty}^{+\infty}\int\cdots\int e^{-z_1^2-z_2^2-\cdots-z_r^2}\,dz_1\,dz_2\cdots dz_r\\
&= e^{r\beta^2/4\alpha} \cdot \alpha^{-r/2},
\end{aligned}$$

from which I find

$$\frac{1}{(\sqrt{\pi})^r}\int\int\cdots\int \xi^i \eta^{2j} \cdot e^{-\xi}\,dt_1\,dt_2\cdots dt_r$$
$$= \left\{\frac{(-1)^i d^{i+2j}\Omega}{d\alpha^i\,d\beta^{2j}}\right\}_{\substack{\alpha\,=\,1\\ \beta\,=\,0}}.$$

The calculations are in no way so complicated that it would be impossible to carry them out completely, especially in view of the fact that we must set β equal to zero. The computations involve the well-known Laplace functions where we need consider only the part free of the variable.

I want very much to verify whether or not the formulas you indicated really prove to be correct.

Sincerely yours,

A. MARKOV

No. 94

(Letter from Markov to Chuprov)

31 January 1917

Highly respected Alexander Alexandrovich:

Your conjecture about the limit of the mathematical expectation of any power of Q^2 is brilliantly verified for the case where the number of occurrences of an event is considered.

My computations, outlined in the previous letter, give the following general expression for the limit of $E\ Q^{2\ell}$:

$$
\frac{1}{(r-\ell)^\ell} \left\{ r(r+2)(r+4) \cdots (r+2\ell-2) - \ell \frac{r}{2}(r+2) \right.
$$

$$
\times (r+4) \cdots (r+2\ell-2) + \frac{\ell(\ell-1)}{1\cdot 2}\cdot\frac{3\cdot 4}{2^2}(r+4)
$$

$$
\cdots (r+2\ell-2) - \frac{\ell(\ell-1)(\ell-2)}{1\cdot 2\cdot 3}\cdot\frac{4\cdot 5\cdot 6}{2^3}(r+6)
$$

$$
\cdots (r+2\ell-2) + \cdots \pm \frac{\ell(\ell-1)\cdots(\ell-k+1)}{1\cdot 2\cdot 3\ldots k}
$$

$$
\times \frac{(k+1)\cdots 2k}{k}(r+2k)\cdots(r+2\ell-2) + \cdots + (-1)^\ell
$$

$$
\left. \times \frac{(\ell+1)(\ell+2)\cdots(2\ell)}{2^\ell} \right\}^2 = \frac{\Phi(r,\ell)}{(r-\ell)^\ell}.
$$

It remains only to prove that the integer-valued function $\Phi(r,\ell)$ of the number r is equal to the product:

$$
(r-1)(r+1)(r+3) \cdots (r+2\ell-3) = \varphi(r,\ell).
$$

But for small ℓ this is not difficult to establish and you have already done so. For successive increases in the number ℓ we consider the two differences

$$\Phi(r + 2, \ell) - \Phi(r, \ell) \quad \text{and} \quad \varphi(r + 2, \ell) - \varphi(r, \ell).$$

It turns out that they reduce to

$$2\ell\, \Phi(r + 2, \ell - 1) \quad \text{and} \quad 2\ell\, \varphi(r + 2, \ell - 1).$$

Consequently, the equality

$$\Phi(r, \ell - 1) = \varphi(r, \ell - 1)$$

allows us to establish the equality

$$\Phi(r + 2, \ell) - \Phi(r, \ell) = \varphi(r + 2, \ell) - \varphi(r, \ell).$$

And hence it follows that the two integral functions

$$\Phi(r, \ell) \quad \text{and} \quad \varphi(r, \ell)$$

can differ only by a constant difference. Then supposing $r = -2\ell + 2$, we obtain two equal numbers

$$\Phi(-2\ell + 2, \ell) = (-1)^\ell \frac{(\ell + 1)(\ell + 2) \cdots 2\ell}{2^\ell}$$

and

$$\Phi(-2\ell + 2, \ell) = (-1)^\ell \cdot 1 \cdot 3 \cdot 5 \ldots (2\ell - 1),$$

for

$$\frac{(\ell + 1)(\ell + 2) \cdots 2\ell}{2^\ell} = \frac{1 \cdot 2 \cdot 3 \ldots \ell(\ell + 1)(\ell + 2) \cdots 2\ell}{2 \cdot 4 \cdot 6 \ldots 2\ell}$$
$$= 1 \cdot 3 \cdot 5 \ldots (2\ell - 1).$$

In this way your assumption can be considered completely proved. Then there remains to put $Q^2 = \dfrac{2}{r - 1}\, x$, and we can establish for x a limit law of probability in the form of the integral $\dfrac{1}{\Gamma\left(\dfrac{r-1}{2}\right)} \displaystyle\int e^{-x} \cdot x^{(r-3)/2}\, dx$. Pearson's formula, if you wish, but very special. Regarding the equality

$$\lim E\, \frac{z^k}{y^k} = \lim \frac{Ez^k}{Ey^k}$$

I have nevertheless not been convinced that the proof of this follows from your reasoning for the case on which I have been working up to now and which, apparently, should remain in the first place in my thoughts.

Sincerely yours,

A. MARKOV

No. 95

(Postcard from Markov to Chuprov)

31 January 1917

Highly respected Alexander Alexandrovich:

Your A, B, C contain the indeterminancy $0/0$. The value of $EQ^4 - 1$ that I have found does not agree with that you have given. I have neither p^{nr} nor q^{nr}, i.e., those terms for which Q reduces to $0/0$.

This circumstance plays an important role in the case when p tends to zero and pn remains unchanged, for in this case q^{nr} cannot be disregarded. Obviously in the case of Q^2 you assume something different than I do. I consider my result correct.

Yours,

MARKOV

No. 96

(Letter from Chuprov to Markov)

1 February 1917

Highly esteemed Andrei Andreevich:

Having received your postcard this morning I began again to do a great deal of thinking about our assumptions for the case where the variable takes one and only one value in all trials and both the numerator and the denominator of Q^2 become zero. The matter as I now see it is fairly clear. My result takes y/y and y^2/y^2 as equal to one, not z/y and z^2/y^2. The discrepancy between these assumptions turns out to be enormous: the assumption that $y/y = 1$[1] leads to $z/y = n$. This does not affect EQ^2 but for higher powers it becomes appreciable.

Consequently, the question arises, what system of assumptions is preferable.

To me it seems natural to take: $y/y = 1$.

Sincerely yours,

A. CHUPROV

---

[1]Chuprov denotes by the symbol y/y the case when the numerator and the denominator of the ratio are simultaneously equal to zero (Editor's note).

No. 97[1]

(Letter from Chuprov to Markov)

1 February 1917

Highly esteemed Andrei Andreevich:

By definition we have:

$$Q^2 = \frac{n(nr-1)}{r-1} \cdot \frac{\sum_{i=1}^{r}(x_{(n)_i} - x_{(nr)})^2}{\sum_{i=1}^{nr}(x_i - x_{(nr)})^2},$$

hence

$$Q^2 = \frac{n^2(nr-1)}{(r-1)} \cdot \frac{\sum_{i=1}^{r}\sum_{j \neq i}[x_{(n)_i} - x_{(n)_j}]^2}{\sum_{i=1}^{nr}\sum_{j \neq i}(x_i - x_j)^2},$$

assuming, with

$$x_1 = x_2 = \cdots = x_{nr}, \frac{x}{y} = 1,$$

we take

$$\frac{(x_i - x_j)^2}{\sum\limits_{i=1}^{nr}\sum\limits_{j \neq i} (x_i - x_j)^2} = \frac{(x_n - x_g)^2}{\sum\limits^{nr}\sum\limits_{i=1 \ j \neq i}(x_i - x_j)^2} = \frac{1}{nr(nr - 1)}.$$

Assuming $z/y = 1$, we take

$$\frac{[x_{(n)ii} - x_{(n)ij}]^2}{\sum\limits_{i=1}^{nr}\sum\limits_{j \neq i} (x_i - x_j)^2} = \frac{1}{n \cdot nr(nr - 1)}.$$

But

$$[x_{(n)ii} - x_{(n)ij}]^2 = \frac{1}{n^2}\left\{\sum\limits_{\eta=1}^{n}(x_{\eta i} - x_{\eta j})^2 \\ + \sum\limits_{\eta=1}^{n}\sum\limits_{g \neq \eta}(x_{\eta i} - x_{\eta j})(x_{gi} - x_{gj})\right\}.$$

Therefore the two assumptions would be compatible if we took

$$\frac{(x_{\eta i} - x_{\eta j})(x_{gi} - x_{gj})}{\sum\limits^{nr}\sum\limits_{i=1 \ j \neq r}(x_i - x_j)^2} = 0.$$

But, assuming $y/y = 1$ and $y^2/y^2 = 1$ simultaneously, we should take

$$\frac{(x_{\eta i} - x_{\eta j})^2 \cdot (x_{gi} - x_{gj})^2}{\left[\sum\limits^{nr}\sum\limits_{i=1 \ j \neq r}(x_i - x_j)\right]^2} = \frac{1}{n^2 r^2 (nr - 1)^2},$$

whence

$$\frac{(x_{\eta i} - x_{\eta j})(x_{gi} - x_{gj})}{\sum\limits^{nr}\sum\limits_{i=1 \ j \neq r}(x_i - x_j)^2} = \frac{(x_{\eta i} - x_{\eta j})^2}{\sum\sum(x_i - x_j)^2} = \frac{1}{nr(nr - 1)}.$$

In a similar manner the simultaneous assumption

$$\frac{z}{y} = 1 \qquad \text{and} \qquad \frac{z^2}{y^2} = 1$$

leads to

$$\frac{(x_{\eta_i i} - x_{\eta_j j})(x_{g_i i} - x_{g_j j})}{\Sigma\Sigma\,(x_i - x_j)^2} = \frac{(x_{\eta_i i} - x_{\eta_j j})^2}{\Sigma\Sigma\,(x_i - x_j)^2} = \frac{1}{nnr(nr - 1)}\,.$$

In this way the assumptions

$$\frac{y}{y} = 1 \qquad \text{and} \qquad \frac{z}{y} = 1$$

are clearly not compatible. The choice between them comes to the choice between equating

$$\frac{(x_i - x_j)^2}{\displaystyle\sum_{i=1}^{nr}\sum_{j\neq i}\,(x_i - x_j)^2} = \frac{1}{nr(nr - 1)} \qquad \text{or} \qquad = \frac{1}{n}\cdot\frac{1}{nr(nr - 1)}\,.$$

It seems to me indisputable that we should prefer $\dfrac{y}{y} = 1$.

Sincerely yours,

A. CHUPROV

---

[1] We have not attempted to correct any of the apparent misprints in the formulas of this letter because in a sense the whole letter is an error. It expresses Chuprov's attitude that when 0/0 occurs in a definition one should attempt to interpret it rather than replace such an erroneous definition by a correct definition, in principle arbitrary (Translators' note).

No. 98

(Letter from Markov to Chuprov)

2 February 1917

Highly respected Alexander Alexandrovich:

For a long time already your attention has been focused on the indeterminancy of the value of EQ^2 that you were considering. This circumstance was clearly observed in my note "On the Coefficient of Dispersion." If Q^2 is not equal to one in exceptional cases, then the equality $EQ^2 = 1$ does not hold either. If, however, we take Q^2 equal to one, then we should also take all powers of it equal to one. Otherwise, all of our computations will take on a fantastic character. As for the ratios y/y and y^2/y^2, it is impossible to derive any kind of results from equating them to one. The appearance of the expression $0/0$ is the big deficiency in all of these computations which you reinforce by breaking up the numerator into parts. I do not know which of your letters dated 1 February should be considered the last one. In one of them you claim that from the equality y/y, it follows that $z/y = n$. Of course it does not follow. But nothing prevents us from taking $Q^2 = n$ in exceptional cases. Only in this case in place of the result that you like, $EQ^2 = 1$, we obtain a different equality

$$EQ^2 = 1 + (n - 1)(p^{nr} + q^{nr}).$$

And if we now take Bortkiewicz's assumption that np remains constant, then the limit for EQ^2 becomes ∞. I do not at all like to argue about the value of $0/0$ when the numerator and denominator are really zero. If the variables x_i can take arbitrarily small values, and you apparently have this case in mind when you refer to Gauss' formula, then it is also possible to carry out the calculations taking $Q^2 = 0$ in the exceptional cases.

Sincerely yours,

A. Markov

No. 99

(Postcard from Markov to Chuprov)

2 February 1917

(z/y) should make $y = 0$ and not something else.

As an addition to the sealed letter I should say that not one of your equalities is a consequence of the assumption that $z/y = 1$. You have to contrive various dubious assumptions, as for example, to take

$$\frac{(x_i - x_j)^4}{(\Sigma\Sigma)^2} = 0$$

identically for all i and j.

It is necessary to eliminate all this and to allow the division of $\dfrac{z}{y}$ into parts only in exceptional cases.

Yours,

A. MARKOV

No. 100

(Letter from Markov to Chuprov)

5 February 1917

Highly respected Alexander Alexandrovich:

The expression for Q^2 contains in the numerator and in the denominator the difference $x_i - x_j$ and takes on a definite meaning in all cases except those where all the x_i are equal. For these exceptional cases Q^2 reduces to 0/0 and a value must be assigned to it arbitrarily since by no kinds of considerations can it be determined. Only after the value of Q^2 is assigned in the exceptional cases can we talk about EQ^2, EQ^4 and so on as having definite values. But in establishing the equality $(z/y)_{y=0} = 1$ we do not determine at the same time the value of $(x_i - x_j)^2/y$ and other more complicated expressions.

In my opinion we can hardly avoid a contradiction if we begin to attribute a definite meaning to all of these expressions, no matter what kind of value we attribute to

$$(z/y)_{y=0}.$$

As for your expressions, things are still worse with them: $E\,\dfrac{(x_i - a_i)^2}{y} = \infty$ since $y = 0$ in the exceptional cases and the difference $x_i - a_i$ is different from zero. Therefore, it is only possible to break up z into parts in the non-exceptional cases to which it is necessary to confine the ratio z/y.

If you put $(z/y)_{y=0} = 1$, the results I gave are obtained; if you put $(z/y)_{y=0} =$ a different value, the result will be different; one can obtain anything.

Sincerely yours,

A. MARKOV

No. 101

(Letter from Markov to Chuprov)

7 February 1917

Highly respected Alexander Alexandrovich:

It is possible to derive the limiting formulas fairly simply after all in the general case also where the variables x_i and $x_{j,i}$ are not Gaussian (according to your terminology) but can have only a definite number of different values (unchanging and with unchanging probabilities). Since all of the s_i increase indefinitely, Laplace's formula can be applied to the variables $z_i = \dfrac{1}{s_i} \Sigma x_{j,i}$ (your notation). For each $x_{j,i}$ let $\alpha^{(1)}, \alpha^{(2)}, \ldots, \alpha^{(n)}$ be possible values with probabilities $p^{(1)}, p^{(2)}, \ldots, p^{(n)}$.

For simplicity we set

$$p^{(1)} \cdot \alpha^{(1)} + p^{(2)} \cdot \alpha^{(2)} + \cdots + p^{(n)} \cdot \alpha^{(n)} = 0,$$

which is not difficult to achieve, replacing $x_{j,i}$ by the difference $x_{j,i} - a$.

In deriving limiting results we restrict ourselves to those cases where, in the totality of all the numbers $x_{j,i}$ the values $\alpha^{(1)}, \alpha^{(2)}, \ldots, \alpha^{(n)}$ appear $p^{(1)} \cdot s + t^{(1)} \cdot \sqrt{s}, \ p^{(2)} \cdot s + t^{(2)} \cdot \sqrt{s}, \ldots, p^{(n)} \cdot s + t^{(n)} \cdot \sqrt{s}$ times, where $t^{(1)}, t^{(2)}, \ldots, t^{(n)}$ are finite numbers and their sum is equal to zero. But in these cases the expression

$$y = \frac{1}{s-1} \sum_{i=1}^{s} \{x_i - x_{(s)}\}^2$$

for sufficiently large values of $s_1, s_2, \ldots, s_r$ will be arbitrarily close to

$$p^{(1)} \cdot \alpha^{(1)} \cdot \alpha^{(1)} + p^{(2)} \cdot \alpha^{(2)} \cdot \alpha^{(2)} + \cdots + p^{(n)} \cdot \alpha^{(n)} \cdot \alpha^{(n)} = A.$$

Turning to the numerator of Q^2, i.e. the expression

$$W = \frac{1}{r-1} \sum s_i (z_i - x_{(s)})^2,$$

we can, on the same basis, put

$$z_i = \xi_i \sqrt{\frac{2A}{s_i}} \quad \text{and}$$

$$x_{(s)} = \sqrt{2A} \left(\frac{\xi_1 \sqrt{s_1}}{s} + \frac{\xi_2 \sqrt{s_2}}{s} + \cdots + \frac{\xi_r \sqrt{r}}{s} \right).$$

In this way the expression for Q^2 reduces to

$$\frac{2}{r-1}\sum s_i \left\{ \frac{\xi_i}{\sqrt{s_i}} - \frac{\xi_1\sqrt{s_1} + \xi_2\sqrt{s_2} + \cdots + \xi_r\sqrt{s_r}}{s_1 + s_2 + \cdots + s_r} \right\}^2 = Q_\infty^2,$$

moreover, $\xi_1, \xi_2, \ldots, \xi_r$ will be independent variables and their probabilities are expressed by the integral

$$\left(\frac{1}{\sqrt{\pi}}\right)^r \int \int \cdots_{\cdots} \int e^{-\xi_1^2 - \xi_2^2 - \cdots - \xi_r^2}\, d\xi_1\, d\xi_2 \cdots d\xi_r.$$

The expression for Q_∞^2 reduces to

$$\frac{2}{r-1}\left\{ \sum \xi_i^2 - \left(\frac{\xi_1\sqrt{s_1} + \xi_2\sqrt{s_2} + \cdots + \xi_r\sqrt{s_r}}{\sqrt{s_1 + s_2 + \cdots + s_r}} \right)^2 \right\}$$

or

$$\frac{2}{r-1}\left\{ \sum \xi_i^2 - \left(\sum \gamma_i \cdot \xi_i \right)^2 \right\},$$

where

$$\gamma_i = \frac{\sqrt{s_i}}{\sqrt{s_1 + s_2 + \cdots + s_r}} \qquad \text{and} \qquad \sum \gamma_i^2 = 1.$$

Then, using the earlier method, we can easily convince ourselves that the mathematical expectations of powers of Q_∞^2 do not depend on the numbers γ_i satisfying the condition $\sum \gamma_i^2 = 1$. Therefore for their computation we can equate all of the γ_i to zero except one which must then be set equal to one.

In this way we are convinced that instead of Q_∞^2 one can consider

$$\frac{2}{r-1}\,(\xi_1^2 + \xi_2^2 + \cdots + \xi_{r-1}^2)$$

which also leads immediately to a well-known result. But it is even clearer that everything is gotten by the consideration that Q_∞^2 is a quadratic form of $r - 1$ variables and for each γ can be represented as the sum $\eta_1^2 + \eta_2^2 + \cdots + \eta_{r-1}^2$.

Once the numbers γ_i satisfy the condition $\Sigma\gamma_i^2 = 1$, it is possible to select, corresponding to them, a system of numbers

$$
\begin{array}{cccc}
\gamma_1^{(1)} & \gamma_2^{(1)} & \cdots & \gamma_r^{(1)} \\
\cdot & \cdot & & \cdot \\
\cdot & \cdot & & \cdot \\
\cdot & \cdot & & \cdot \\
\gamma_1^{(r-1)} & \gamma_2^{(r-1)} & \cdots & \gamma_r^{(r-1)}
\end{array}
$$

in such a way that the expressions

$$
\eta_1 = \gamma_1^{(1)} \cdot \xi_1 + \gamma_2^{(1)} \cdot \xi_2 + \cdots + \gamma_r^{(1)} \cdot \xi_r
$$

$$
\eta_{r-1} = \gamma_1^{(r-1)} \cdot \xi_1 + \gamma_2^{(r-1)} \cdot \xi_2 + \cdots \gamma_r^{(r-1)} \cdot \xi_r
$$
$$
\eta_r = \gamma_1 \cdot \xi_1 + \gamma_2 \cdot \xi_2 + \cdots + \gamma_r \cdot \xi_r
$$

will satisfy the condition

$$
\eta_1^2 + \eta_2^2 + \cdots + \eta_r^2 = \xi_1^2 + \xi_2^2 + \cdots + \xi_r^2
$$

and

$$
\begin{vmatrix}
\gamma_1^{(1)} & \gamma_2^{(1)} & \cdots & \gamma_r^{(1)} \\
\cdot & \cdot & \cdots & \cdot \\
\cdot & \cdot & \cdots & \cdot \\
\gamma_1^{(r-1)} & \gamma_2^{(r-1)} & \cdots & \gamma_r^{(r-1)} \\
\gamma_1 & \gamma_2 & \cdots & \gamma_r
\end{vmatrix} = 1.
$$

Then the transition from $\xi_1, \xi_2, \ldots, \xi_r$ to $\eta_1, \eta_2, \ldots, \eta_r$ reduces Q_∞^2 to the form

$$
\frac{2}{r-1}\left\{\sum \eta_i^2 - \eta_r^2\right\} = \frac{2}{r-1}\{\eta_1^2 + \eta_2^2 + \cdots + \eta_{r-1}^2\}
$$

and at the same time

$$
e^{-\xi_1^2 - \xi_2^2 - \cdots \xi_r^2} \, d\xi_1 \, d\xi_2 \cdots d\xi_r
$$

is replaced by

$$
e^{-\eta_1^2 - \eta_2^2 - \cdots \eta_r^2} \, d\eta_1 \, d\eta_2 \cdots d\eta_r.
$$

Subsequently, η_r drops out. In this way everything is explained fully and it becomes possible to consider the probability instead of the mathematical expectations. The result is inapplicable to the case when the number of different values of x is infinitely large.

Sincerely yours,

A. MARKOV

No. 102

(Letter from Chuprov to Markov)

8 February 1917

Highly esteemed Andrei Andreevich:

"The secret of zero divided by zero" that envelops the question about EQ^{2k} greatly intrigues me; better to say troubles me. Unfortunately I do not have the opportunity to ponder it intently since these days are taken up with pressing work. The Central Statistical Committee is advancing a new grandiose project "of reorganization of the statistical department in the Empire". The scope is very wide, enormous appropriations, new staff, advancement of the Committee in official rank to Central Administration, and not only is there no benefit from this pursuit but rather, in my opinion, damage is foreseen. Our Central Committee has gradually sunk to an unimaginably low level; its work is beneath all criticism. It is explained mainly by the fact that from a semi-scientific institution as it should have been it has turned into some kind of statistical department—into an office which manufactures statistical answers written for form only, instead of carrying out statistical investigations. And as long as the bureaucratic spirit of "service" that rules there is not replaced by the spirit of scientific service to knowledge of Russia, no kind of grants, no kind of staff will help anything. One cannot expect good from an infusion of new wine into old wine skins.

And I consider it my duty to express this point of view. Of course I do not expect any special success but to compel this Committee to rouse themselves a bit and it seems that raising the basic question with thorough abruptness will impel them to start to think about how it is necessary to begin the reform. This must be done, even if for conscience' sake, but there is a lot of bother with such work and the bother is extremely unpleasant: it is disgusting to be investigating slovenly work with the sole aim of portraying with complete intelligibility every degree of its impropriety.

As for the coefficient of dispersion, it apparently has to be acknowledged reluctantly that one cannot avoid contradictions if in the exceptional cases one assumes determinate values in breaking up 0/0 into parts.

In this way that part of my work where I change over from E to the sums is eliminated. But does it really have to be assumed that everything else also should go to pot? It is distressing to reconcile oneself to this. True, there is the consolation that the work was not completely futile since it caused you to study these questions and you secured the main results.

It is curious that my basic approach not only leads to the same results for the limiting values, but also imposes the same essential condition on the distribution law of the random variables as your new approach; that is, that Laplace's formula could be applied in the limit to the arithmetic mean (that, as s increases, the values of $\dfrac{1}{s} \cdot \dfrac{\mu_4}{\mu_2^2}$, etc. tend to zero).

Another restriction imposed by your result (that the variable could assume only a finite number of different values and did not vary continuously) in all probability will somehow manage to be removed. Your new result evades the question of the exceptional cases, limiting itself to observing that their probability of occurrence is vanishingly small. They can in fact be evaded on this ground and for other reasons: for sufficiently large s, p^s is also arbitrarily close to zero. Bohlmann, in his work concerning the conditions under which the replacement of the mathematical expectation of a function by a function of the mathematical expectation is allowed (*Math. Annalen*, 1913), even more decisively proposes not to consider at all the exceptional cases and, eliminating them, to limit his consideration to Hauptwerte, in his notation. Perhaps this outcome ought also to be recognized as a way out.

My original scheme of proof attempted to get around the obstacle in a somewhat different way. I computed, with k a positive integer, Ey^{k+1} and Ezy^k and showed that both expressions were identical and then proved that if $Ezy^k = Ey^{k+1}$ for $k = 0, 1, 2, \ldots$, then $E\dfrac{x}{y} = 1$. It is annoying that although the manuscript had been given to the editor of *Proceedings of the Polytechnic Institute* by the middle of September, up to now I have not received even the beginnings of the proofs and I do not remember right now exactly how the exceptional cases are put in there.

Sincerely yours,

A. Chuprov

No. 103

(Letter from Markov to Chuprov)

10 February 1917

Highly respected Alexander Alexandrovich:

In my opinion all of your calculations can maintain their validity; only the cases when the denominator of Q^2 is equal to zero should be singled out. For them it is impossible to divide Q^2 up into the terms you indicated but it is necessary somehow to assign a value to Q^2 here and then powers of Q^2 should take on corresponding values. In the normal case the probabilities of the exceptional cases are small and tend to zero as n increases. Therefore, for the limiting results it is of no importance how Q^2 is assigned as long as this value does not increase too fast. But for the Bortkiewicz-Poisson case either p^n or q^n is no longer an infinitely small quantity. Therefore for such a case the choice of the value of Q^2 in the exceptional cases already plays an essential role.

My results have come to very little inasmuch as a large part of them proved to be superfluous once it became clear that the numerator of Q^2 can be represented in the form[1]

$$\overline{X}_1^2 + \overline{X}_2^2 + \cdots + \overline{X}_r^2,$$

where $\overline{X}_1, \overline{X}_2, \ldots, \overline{X}_r$ are linearly related to

$$z_i = \frac{1}{s_i} \sum x_{j,i}.$$

Will you not try to apply Pearson's formulas to the following mathematical expectations:

$$E\, U = 0, \quad E\, U^2 = \left(\frac{1}{2}\right)^2, \quad E\, U^3 = 0, \quad E\, U^4 = \left(\frac{1 \cdot 3}{2^2}\right)^2,$$

$$E\, U^5 = 0, \quad E\, U^6 = \left(\frac{1 \cdot 3 \cdot 5}{2^3}\right)^2, \ldots ?$$

Sincerely yours,

A. MARKOV

---

[1]Here the mathematical notation, in particular $\overline{X}_1^2$ and others is not clear since their meaning is apparently explained in letters of Chuprov which are not at our disposal (Editor's note).

No. 104

(Postcard from Markov to Chuprov)

13 February 1917

Highly respected Alexander Alexandrovich:

The mathematical expectations that I reported are entirely possible since they correspond to the product of two independent variables of the Gaussian (as you put it) type. But the formulas you cite, as far as I understood them, should give quite inappropriate results for the mathematical expectations of powers higher than the fifth.

Sincerely yours,

A. MARKOV

No. 105

(letter from Markov to Chuprov)

27 February 1917

Highly respected Alexander Alexandrovich:

Your work frightens me with its abundance of complicated calculations which are difficult for me to look through in all their plentitude, all the more since I have glaucoma in one eye. Therefore all of my remarks will be based only on a superficial review. First of all, regarding the name of one of Gauss' formulas, I think that it should be combined with the names DeMoivre and Laplace.

As for the question about the applicability of this formula, it is impossible to resolve it in a positive sense by considering certain mathematical expectations. Since Q^2 is not the usual arithmetic mean for which Laplace's formula (in fact Laplace and not Gauss) is established, there is no firm basis for applying this formula to Q^2. And as soon as you prove that the limit of $(Q^2 - 1)^3$ is not zero, it can be asserted with some basis that it is impossible to apply Laplace's limit formula to Q^2. But even here it is impossible to express oneself decisively in view of the fact that the mathematical expectations can be determined by infinitely distant elements, which holds, for example, in Liapunov's cases which I have been thinking about for a long time. Infinitely improbable, infinitely distant values can change the mathematical expectations as much as you please. The necessity of the transition to Pearson's formulas is even more questionable.

At the same time I should remark that I do not consider it possible to talk about a distribution curve for $n = \infty$. One can only talk about the limiting probability for some kind of interval but not about the curve. In the normal case there is no kind of curve—not in the limit, but there are only separate points, not even in the limit.

Regarding the formula of Pearson that you refer to, not only do the mathematical expectations of the powers considered by you correspond to it but also others, and until you have proved that they agree with it for your case it is impossible to assert its applicability. It would be possible to obtain a weak basis for such an assertion by computing the mathematical expectation of the next power after those that are used only for determining the constants. In any case the calculation of the limit values of the mathematical expectation of Q^6 and Q^8, even under certain restrictions, interests me if the simplest problem about identical sequences of independent trials comes in under these conditions. This simplest case appears to me to be the most important also and, in my opinion, it would be very good if you would dwell on it especially. The case of p approaching zero with np = constant does not seem particularly important to me since with such a change in n and p one must constantly change the problem. By the way, I should tell you that based on your opinion of Novoselsky's[1] work "Mortality . . .", he should not be given the prize and I have said so in the general meeting of the Academy. I do not dispute your opinion but you did not point out its special merits; in my opinion there are none. I am enclosing the proofs of my short note which will appear soon.

Sincerely yours,

A. MARKOV

---

[1] S. A. Novoselsky—a prominent Russian statistician. In the 20s and 30s he was the director of the department of social statistics of the Leningrad regional plan where tables of the mean life expectancy and mortality rates of the population were compiled.

A Review of the Correspondence Between A. A. Markov and A. A. Chuprov

by

Kh. O. Ondar

Translators' note: This is a somewhat shortened version of the editor's review. The omissions are principally of two kinds. Extensive quotations from the Markov-Chuprov correspondence have been replaced by references to the appropriate point in the text. In addition, Ondar's detailed treatment of some of the formulas related to Lexis's dispersion theory has been omitted. A thorough discussion of this theory and its historical background is given in Chapter 3 of C. C. Heyde and E. Seneta's book, *I. J. Bienaymé, Statistical Theory Anticipated* (Springer-Verlag, 1977). Omissions have ordinarily been indicated by ellipses, except for the quotations, where it is clear from the context that they have been omitted. The editor's interesting and useful comments concerning the influence of the correspondence on Markov and Chuprov and on their later work have been retained.

In the correspondence of A. A. Markov and A. A. Chuprov basic attention is given to the work of V. Lexis on dispersion and stability of statistical series. For this reason let us briefly stop to consider the nature of this study and the way it arose.

The historical reason for Lexis's study of the theory of measurement of stability of statistical series was an argument over the incorrectness of conclusions drawn by followers of the Belgian statistician Adolphe Quetelet from his publications on social statistics, although his opponents on their side resorted to even less legitimate arguments. The essential point of this argument was the philosophical problem of "free will."

Already at the beginning of the last century Quetelet and the French statistician Guerry had established that suicide, divorce, and also various crimes were repeated from year to year with definite stability. From this the followers of Quetelet concluded that each person has a unique tendency to crime, that to a person it only seems that he could behave one way or another and really his will is not free, that is, it is conditioned by the effect of his environment.

The incorrectness of the point of view of Quetelet's followers brought down on itself the no less incorrect argument of their opponents who, because they did not know the basic laws of probability theory, could not conduct the argument on the essence of the problem but rather appealed to a feeling of outrage to human dignity. In this way their judgement of statistical stability had a purely subjective nature.

The scientific solution of this argument was provided by the German scientist V. Lexis. Contrary to the ruling ideas of his time, Lexis energetically advocated the necessity of developing statistical theory based on the theory of probability. He proposed a soundly based solution of the problem of stability and found an objective measure for it. In 1877 Lexis expounded the basic ideas of his theory in the work "Zur Theorie der Massenerscheinungen in der menschlichen Gesellschaft."

Lexis constructed the theory of stability in connection with relative frequencies and sixteen years later V. I. Bortkiewicz extended his construction to mean values. Lexis's basic idea was that the real fluctuations occurring among observations are the same as the fluctuations under the conditions of Jacob Bernoulli's theorem: 1) constant probabilities, and 2) independence of the individual observations on which the observed frequencies and the mean values making up the terms of the statistical series are based.

As a criterion for estimating the stability of statistical series, Lexis suggested the value of the so-called dispersion coefficient Q, computed from the formula

$$Q = \sqrt{\frac{\frac{1}{\mu} \sum_{i=1}^{\mu} (\gamma_i - \gamma)^2}{\frac{pq}{n}}}$$

where μ is the number of series observed with n trials in each, γ_i is the relative frequency of occurrence of the event in question in the ith series of n trials. γ is the relative frequency of the event in all μ series (a total of μn trials), $p = E\gamma$ is the mathematical expectation of the relative frequency in all μn trials and $q = 1 - p$.

If Q comes out close to 1, the dispersion of the series is considered to be essentially normal (in the sense of Lexis). If $Q > 1$ we shall say, following Lexis, that the dispersion of the series is supernormal (and correspondingly the stability of the series is subnormal). On the other hand, if

$Q < 1$ the dispersion of the series is subnormal (and correspondingly the stability of the series is supernormal).

In the correspondence between Markov and Chuprov the difference between their approaches to the explanation of supernormal and subnormal dispersion is very interesting. While Markov insistently emphasized the dependence of the dispersion on differences among the probabilities lying at the basis of the constructed series of trials, Chuprov put in the forefront the presence of positive or negative dependence among the observations.

. . .

How highly Chuprov valued Markov's formula (the final formula in letter no. 16) is clear from the quotation given below. In letter no. 24 dated 21 November 1910 he wrote: "The enormous interest of [your] formulas is obvious. In the field of dispersion theory I think this represents the greatest possible generality in the formulation of the problem and the most striking clarity of solution. It seems to be impossible to go any further in this direction. On the basis of your formula (also bringing in Bohlmann's formula for a parallel clarification from a somewhat different point of view) it is now possible to cover the whole theory of stability in a completely finished form, to sketch freely its entire construction without the yoke of the fortuitousness of its historical growth."

And indeed, the unification of these two approaches to the clarification of various kinds of stability was accomplished later by Chuprov himself in his paper "Zur Theorie des Stabilität statistischer Reihen" in the *Skandinavisk Aktuarietidskrift*[1].

In connection with the problem of dispersion, the question arose of applicability of the law of large numbers to various cases of positive dependence among trials. It is appropriate to remark that nowhere in the correspondence is any attention paid to the distinction between the law of large numbers and the central limit theorem.

For example, in letter no. 60, dated 31 October 1913, Markov writes " . . . but to make up for it I shall also include, under the law of large numbers, the expression for the probability by the well-known integral." Until the work of Markov, Liapunov was concerned with the law of large numbers, not the central limit theorem. For Chuprov also, the law of large numbers was inseparable from the central limit theorem. This is clear from the following words of his: "The derivation of the law of large numbers in the form given to it by Bernoulli and Laplace rests on the assumption of invariability of the underlying conditions and the independence of the individual trials."[2]

Having made this observation we want to emphasize that Markov and Chuprov carried on a correspondence that touched upon burning questions of the theory of probability and mathematical statistics so that they were concerned primarily with the perspective of further development and generalization of the law of large numbers and the central limit theorem to cases that had not been studied at all up to that time.

By special examples, questions were raised, not only about the applicability of the law of large numbers but also about the application of the central limit theorem to cases of dependent trials.

. . .

Translators' note: The editor, Ondar, cites the example introduced by Markov in letter no. 7, dated 15 November 1910, giving some details of these computations and related work. Ondar emphasizes that this example is of great interest because of its historical importance.

Above all, Markov's example distinctly shows that the dispersion in the sense of Lexis can be normal even for dependent trials. In addition, in 1932 at the International Congress of Mathematicians in Zurich, Academician S. N. Bernstein delivered a paper "On dependence among random variables."[3] In this paper Bernstein introduced, in particular, the "random bridge"[4] generalizing the notion of "Markov chain."

. . .

Markov's assertion in letter no. 9 dated 17 November 1910 that for his example the departure of the dispersion from normal is revealed only by consideration of higher order moments turned out to be incorrect. (*Translators' note:* here Ondar quotes from Markov's letter no. 9).

In this way, clarifying the question of the preservation of the form of the normal distribution in the case where Lexis's $Q = 1$, < 1, or > 1, Markov and Chuprov in various special examples studied the extension of the central limit theorem to the case of weakly dependent trials.[5]

In 1913 the 200th anniversary of the publication of Jacob Bernoulli's *Ars Conjectandi* took place. The two authors agreed to take on themselves the responsibility for organizing a ceremonial meeting of the Academy of Sciences devoted to this important date. At this meeting they gave interesting papers which were later published.[6]

Great attention was paid in the correspondence to the "law of small numbers" developed by Bortkiewicz in 1898. This "law" did not lead to sharp agruments and in the end the two authors agreed that Lexis's criterion Q was not applicable to this case.

In the correspondence the scientific merit of Bohlmann, Pearson and other mathematicians is evaluated, and some questions of priority are also touched upon. At first Markov did not acknowledge the merit of Bohlmann or Pearson at all. This is clear from his words: " . . . it is clear to me that neither Bruns nor Nekrasov nor Pearson has done anything worthy of note" (letter no. 3 dated 6 November 1910) " . . . I cannot attach any importance to the concoctions of Pearson and others establishing some kind of asymmetric expressions." (letter no. 9 dated 17 November 1910)

It is clear from letter no. 7 dated 15 November 1910 that the works of Bohlmann were not known to Markov before the time that Chuprov called his attention to them. Markov indicated to Chuprov a certain disbelief that Bohlmann had been working on the question of dispersion from the point of view of dependence among the observations.

However, gradually, under Chuprov's influence, Markov changed his opinion of the work of the scientists I have mentioned: "I am prepared to admit that Bohlmann gave an elegant special formula" (postcard no. 12 dated 18 November 1910). In letter no. 43 dated 5 December 1910 he writes, "I most humbly beg you, if it won't be too much trouble, to pass on the two articles I have sent you to Bohlmann when you are abroad."

Markov's initial disdain for Bohlmann's work, "Die Grundbegriffe der Wahrscheinlichkeitsrechnung in ihrer Anwendung auf die Lebensversicherung" (1909), where, in particular, Bohlmann's formula in the first footnote to letter no. 10 was obtained, is apparently explained by the fact that it was based on an axiomatic construction of the theory of probability. Indeed, as is well known, Markov "had a negative attitude toward axiomatics."[7]

The evolution of Markov's attitudes was especially marked in the case of Pearson. His original complete disdain for Pearson later changed to some interest in him. This is clear from the following statements of Markov. ". . . Pearson's formulas become more or less successful empirical formulas" (letter no. 49, dated 1 December 1912). " . . . I plan to send the messenger from the Academy of Sciences to you on Monday between three and four o'clock for *Biometrika*. I assume that the first five volumes are quite enough for me since in the *Calculus of Probability* I shall have to speak about Pearson" (letter no. 53, dated 8 December 1912). "The question about Pearson's curves interests me in the following way: whether fitting them is a well-defined operation and whether they really reproduce the observations well in the examples cited" (letter no. 65, dated 1 February 1916). "I consider comparisons with Pearson's formulas interesting and am very glad that you are doing this. It represents a theoretical example on which the suitability of these formulas can be tested" (letter no. 91, dated 29 January 1917).

Finally, in letter no. 94, dated 31 January 1917, Markov communicated a concrete result related to Pearson's χ^2-criterion of goodness of fit.

Looking at the random variable

$$x = \frac{r-1}{2} Q^2,$$

where r is the number of trials, Markov established that this random variable has a limiting distribution of the form

$$\frac{1}{\Gamma\left(\dfrac{r-1}{2}\right)} \int_{o}^{x} e^{-x} x^{(r-3)/2} \, dx. \tag{1}$$

Let $G(x)$ be the distribution function of Pearson's goodness of fit criterion χ^2.

As is well known, then

$$
G(x) = \begin{cases} 0 & \text{if } x \leqslant 0, \\[2ex] \dfrac{1}{2^\lambda \Gamma(\lambda)} \displaystyle\int_0^x y^{\lambda-1} e^{-y/2}\, dy, & \text{if } x > 0, \end{cases}
$$

where $\lambda = \dfrac{k}{2}$ and k is a natural number which is called the number of degrees of freedom.

Now if we substitute $x = \dfrac{y}{2}$ in formula (1) we obtain

$$
\frac{1}{\Gamma\left(\dfrac{r-1}{2}\right)} \int_0^x e^{-x} x^{(r-3)/2}\, dx = \frac{1}{\Gamma\left(\dfrac{r-1}{2}\right)} \int_0^y e^{-y/2} \left(\frac{y}{2}\right)^{(r-3)/2} \cdot \frac{1}{2}\, dy
$$

$$
= \frac{1}{2^{(r-1)/2}\, \Gamma\left(\dfrac{r-1}{2}\right)} \int_0^y e^{-y/2}\, y^{(r-1)/2 - 1}\, dy.
$$

Thus Markov considered the case where $\lambda = \dfrac{r-1}{2}$ and consequently $k = r-1$, that is, the number of degrees of freedom is one less than the number of trials.

In this way it was no accident that Markov, studying the distribution law of Q^2, arrived at the same distribution law that corresponds to the special case of χ^2 when the number of degrees of freedom is one less than the number of observations.

Observing the connection between Lexis's criterion Q and Pearson's χ^2 criterion of goodness of fit, R. A. Fisher wrote: "It is of interest to note that the measure of dispersion, Q, introduced by the German economist Lexis, is, if accurately calculated, equivalent to χ^2/ℓ of our notation. In the many references in English to the method of Lexis, it has not, I believe, been noted that the discovery of the distribution of χ^2 in reality completed the method of Lexis. If it were desired to use Lexis's notation, our table could be transformed into a table of Q merely by dividing each entry by ℓ."[8]

This connection is clarified in R. K. Bauer's article "Die Lexissche Dispersionstheorie in ihren Beziehungen zur modernen statistischen Methodenlehre, insbesondere zur Streungsanalise" (*Mitteilungsblatt für mathematische Statistik und ihre Anwendungsgebiete*, 1955, no. 7) where he obviously reproduces Fisher's deduction of the connection between Lexis's Q^2 and χ^2. There Bauer introduces the following considerations. Looked at in connection with the theory of dispersion, Pearson's χ^2 method is a variant of Lexis's Q^2. As is known, χ^2 for frequencies of trials is defined as the sum of ratios of squares of deviations of observed from expected values to the expected values. If one starts from the consideration of the formula for χ^2 with r groups of alternative events with observed frequencies np_i and constant expected value np, then, taking account of the identity $p + q = 1$ and the identity that serves to define Q^2, we obtain

$$\chi^2 = \sum_{i=1}^{r} \left\{ \frac{(np_i - np)^2}{np} + \frac{(nq_i - nq)^2}{nq} \right\} = \sum_{i=1}^{r} \frac{(np_i - np)^2}{npq}$$

$$= \frac{n \sum_{i=1}^{r} (p_i - p)^2}{pq} = rQ^2.$$

It is appropriate to observe that Lexis's idea did not only lie at the base of the χ^2-criterion, but also the basic idea led to the development of the analysis of variance, the method of correlation, and other ideas.

The efforts to extend the distribution of Lexis's criterion to the case of many random variables deserves special mention. The first attempts in this direction were already made by Chuprov[9] and the subject received its later development in the works of J. Wishart,[10] S. Wilks,[11] and others.

Looking at the works of English statisticians in 1918, Chuprov wrote: "The most important work in the field of theoretical statistics at the present time is being carried out in England.

"In the present decade Edgeworth, Pearson with his many students, and others have developed extremely interesting work, which I value highly. However, this work is cloaked in mathematical attire that makes it unattractive to continental investigators accustomed to more rigor in proofs. Let me cite an example: a great Russian scholar in the field of probability, Markov, has confessed that he is completely unable to overcome his aversion to Pearson's mathematical reasoning. And I know many colleagues who, like Markov, put the English researchers into the category of the unread."[12]

The examples of successful application of Pearson's formulas pointed out by Chuprov could not fail to excite Markov's curiosity and to lead him to try to find a rational justification for Pearson's constructions[13]. Such an influence on Markov is also shown in his correspondence with E. E. Slutsky in connection with Slutsky's book, *The Theory of Correlation and Elements of the Study of Distribution Curves,* published in Kiev in 1912.

In his letter to Chuprov of 26 November 1912 (no. 47), Markov wrote, "I am corresponding with E. E. Slutsky in connection with his book which interests me although I do not like it very much."

Unfortunately, we have been able to find only one letter of Slutsky's, ardently defending the work of the English biometricians. Let us introduce some excerpts from this interesting letter:

"Highly esteemed Andrei Andreevich,

I believe that you consider my little book incomprehensible for the same reasons for which you find Pearson's work antimathematical. Of course it is a question of the inadequate mathematical rigor of the basic foundation of the theory and methods.

Despite my deep and ardent admiration for your knowledge and authority, I permit myself to have a contrary opinion.

I believe that the inadequacies of Pearson's exposition are temporary, of the same sort as those occurring historically in the mathematics of the seventeenth and eighteenth centuries. A strong foundation under this work of geniuses was built only post factum. The same will happen with Pearson."[14]

Besides the major problems, already indicated, in the application of the theory of probability, a series of substantive problems connected with Lexis's theory of dispersion were raised in the correspondence between Markov and Chuprov. In applying the method of mathematical expectations to the powers of Q^2, it was necessary to resolve the general problem of finding the mathematical expectation of a fraction whose numerator and denominator are mutually dependent random variables. This problem was considered later by Chuprov in his paper "On the mathematical expectation of the quotient of two mutually dependent random variables" (*Works of Russian Scholars Abroad*, v. I, Berlin, 1921).

In the correspondence between Markov and Chuprov much attention and laborious computation were devoted to the question of the probable error of Q^2 which did not yield a precise and complete solution. Thus it was necessary to resort to approximate computation of bounds for the true value of this error. The solution given by Bortkiewicz in his work "Über den Präzisionsgrad des Divergenzkoeffizienten" (*Mitt. des Verbandes der öster. und ungar. Versicherungtechniker*, H. V., Wien, 1901) could not satisfy any scholar who was accustomed to significantly more rigor in his inferences.

This is far from an exhaustive enumeration of the questions that arose in the correspondence between A. A. Markov and A. A. Chuprov. Without doubt this correspondence greatly influenced their work. Three articles of Markov devoted to the coefficient of dispersion[15] are clearly associated with it. The correspondence also had an influence on his later works[16].

Moreover, after the first year of his correspondence with Chuprov, in letter no. 53 of 8 December 1912, Markov wrote to him: "Our previous correspondence had some influence on the first chapter of my book: I dwell more on the clarification of the fundamental concepts, but of course I do not change the direction I have taken."

Directly associated with the correspondence is the work of Chuprov cited earlier, "On the mathematical expectation of the quotient of two mutually dependent random variables," where cardinal questions of the theory of correlation were solved (the mathematical expectation of the correlation coefficient and its systematic error).

B. I. Karpenko, a student of Chuprov, remarked, "It is especially impossible not to mention the intensive correspondence of A. A. Chuprov with A. A. Markov which lasted several years and contributed to A. A. Chuprov's mastery of the method of moments."[17]

A. A. Chuprov's work "Zur Theorie der Stabilität Statistischer Reihen" (1918–1919), exhausting the theme of dispersion (in Lexis's formulation) marked the completion of the solution of problems which, in their essentials, were raised in the correspondence of A. A. Chuprov with A. A. Markov. At the same time this work contains a critical analysis of the weak points of Lexis's theory of dispersion.

---

[1]Russian translation: *Questions of Statistics*, Moscow, Gosstatizdat, 1960, pp. 222–270.

[2]Chuprov, A. A., *Essays on the Theory of Statistics*, 3rd edition, Moscow, Gosstatizdat, 1959, p. 200.

[3]Bernstein, S. N., *The Present State of the Theory of Probability*, GTTI, 1933.

[4]It is appropriate to remark that the case of a "returning chain" (random bridge) was also considered before Bernstein's work. Already in 1927 E. E. Slutsky called a sequence "connected" when its terms are mutually correlated, and he looked at the special case of a symmetric distribution of the correlational connections of each term of the series with its preceding and following values. (*Voprosy Kon'junktury*, 1927, vol. III, no. 1; see also Slutsky, E. E., *Selected Works*, Moscow, Press of the Academy of Sciences of the U. S. S. R., 1960, p. 101).

[5]This result was obtained in the general case by S. N. Bernstein in 1926 (*Math. Annalen*, 1926, vol. 97, pp. 1–59).

[6]Markov, A. A., Appendix 3 of this book. Chuprov, A. A., Appendix 4 of this book.

[7]Markov, A. A., *The Calculus of Probability*, 4th Edition, Moscow, 1924, p. XIII.

[8]Fisher, R. A., *Statistical Methods for Research Workers*, 13th Edition, Hafner Publishing Company, Inc., New York, 1958, p. 80.

[9]Tschuprow, Al. A., "Uber normal stabile Korrelation," *Skandinavisk Aktuarietidskrift*, 1923.

[10]Wishart, J., "The Generalised Product Moment Distribution in Samples from a Normal Multivariate Population," *Biometrika*, 1928, v. 20A.

[11]Wilks, S. S., "Certain Generalisations in the Analysis of Variance," *Biometrika*, 1932, v. 24.

[12]Chuprov, A. A., "The theory of stability in statistical series," *Questions of Statistics*, Moscow, Gosstatizdat, 1960, p. 226.

[13]A theoretical basis of Pearson's curves is given by A. A. Markov in the paper "On some formulas of the calculus of probability," *Izvestiia, Akademii Nauk*, Series VI, 1917, v. II.

[14]Archives of the Academy of Sciences, Reserve 173, inventory 1, No. 18.

[15]"On the coefficient of dispersion," *Izvestiia, Akademii Nauk*, Series VI, 1916, v. 10, No. 9, pp. 709–718; "On the coefficient of dispersion of small numbers," *Strahovoe Obozrenie*, 1916, No. 2, pp. 55–59. "On the coefficient of dispersion (second note)," *Izvestiia, Akademii Nauk*, series VI, 1920, v. 14, No. 1–18, pp. 191–198.

[16]See, for example, "An example of statistical research on the text of *Eugene Onegin,* illustrating the dependence of trials in a chain," *Izvestiia, Akademii Nauk,* series VI, 1913, v. 7, pp. 153–162; "On an application of statistical method," *Izvestiia, Akademii Nauk,* series VI, 1916, v. 10, No. 4, pp. 239–242; and Appendices 1 and 3 of this book.

[17]Karpenko, B. I., *The Life and Work of A. A. Chuprov — Scientific Writings on Statistics,* v. 3, Moscow, Press of the Academy of Sciences of the U. S. S. R., 1957, p. 298.

Appendices:
A. A. Markov and A. A. Chuprov
on the Law of Large Numbers

On the Basic Principles of the Calculus of Probability and on the Law of Large Numbers[1]

by

A. A. Markov

This note was prompted by A. A. Chuprov's book *Essays on the Theory of Statistics*[2]. Although two people did write about the *Essays* in the *Journal of the Ministry of Public Education* in 1910, neither of them dealt appropriately with the most important point, that in the lively and interesting third and fourth essays there is an incorrect attitude toward the basic principles of the calculus of probability and the law of large numbers that grows out of this calculus.

Having reason to fear that such an attitude may spread among statisticians, I consider it my duty to explain its incorrectness for those, of course, who wish to understand me. It is appropriate to begin with the notion of probability itself.

The calculus of probability itself can be discussed without much regard to the question of subjectivity, objectivity or reality of probabilities, but it cannot take a step without the probabilities of events on individual trials, given as numbers corresponding to the conditions under which the trials are carried out. The calculus of probability is concerned with these conditions only insofar as they lead to the determination and variation of these probabilities.

I shall not stop to consider the hopeless question of what the circumstances should be in order for these probabilities to be determined but I must emphasize the fact that the computation of probabilities always supposes certain conditions. Turning to the celebrated theorem ordinarily called the law of large numbers, I should remark that in it the probabilities of some events are taken to be given and the probabilities of others are computed on the basis of the laws of addition and multiplication of probabilities.

I shall not defend these basic theorems connected to the basic notions

of the calculus of probability, notions of equal probability, of independence of events, and so on, since I know that one can argue endlessly on
the basic principles even of a precise science such as geometry; if we consider it possible to verify its conclusions by real measurements then we
must also, for example, admit that the ratio of the diagonal of a square to
its side is different for different squares and is never equal to $\sqrt{2}$.

One who does not want to know the basic principles of one science or
another can calmly reject all of its conclusions insofar as he himself will
not depend on them. However, one who wants to use the conclusions of
a science, applying them to real facts, cannot reject its basic principles
and not give a certain degree of reality to its objects without falling into
contradictions even with himself.

The basic object of the calculus of probability, as I have already emphasized, is the probability of events on separate trials; without this probability there is no law of large numbers. However, in the *Essays* we find a
persistent attempt to establish that such a probability does not exist and
on page 203 it is stated directly that "it is advisable to recognize, without
playing with words, that neither probability nor the law of large numbers
has anything to do with an isolated case."

If, immediately after these words A. A. Chuprov had attempted to
express the law of large numbers even in the formulation given in my book
The Calculus of Probability and which, in his words (p. 199, note 3)[3],
represents only part of the law of large numbers, not containing it completely, then he would immediately have fallen into contradiction with
himself.

In order to clarify this matter, let us take the formulation of the law of
large numbers from my book, though it be only part, according to the
words of Chuprov, and the whole must be sought later in the *Essays*, one
of which is called "The law of large numbers": "For a sufficiently large
number of independent trials the probability is arbitrarily close to certainty that the ratio of the number of occurrences of the event to the number of trials will be arbitrarily close to the arithmetic mean of the probabilities of the events".

This is a simple and quite remarkable theorem of the calculus of probability. The arithmetic mean of the probabilities of the events on the separate trials plays an important role in it. If all the probabilities are different, and this possibility is allowed by the theorem itself, then each
probability is associated with just a single case. Of course the probability
is connected with an individual case even in Bernoulli's theorem, but
there the connection is to a certain extent masked by the equality of the
probability for all trials.

Without connecting the probabilities to individual trials it is impossible
to formulate either the law of large numbers or the simple theorem of
Bernoulli.

Now we must turn our attention to the fact that in the theorem we deal with independent[4] trials. The notion of independence was established in the calculus of probability in conformity with the ordinary notion of independence and with problems of this calculus which connects all of its definitions with the notion of probability.

In the *Essays* the notion of independence is taken, in the words of the author, as a technical term of the calculus of probability (page 167), but at the same time this is found to be eliminated.

I shall not expand on the fact that it is eliminated together with the notion of probability associated with individual cases since the necessity of this probability for the calculus of probability has already been clarified sufficiently. But I must emphasize that it is rendered useless by the absence from the *Essays* of that clarity and definiteness that the calculus of probability requires.

The lack of definiteness is already shown by the fact that all of the problems in the *Essays* are connected with urns and lots, even double.

Urns and lots do not change the notion of probability but only illustrate it and illustrate it quite poorly if they concentrate our attention not on what is given but rather on what remains unknown. They can be accepted as illustrations of problems of the calculus of probability only when there is no doubt that the designation of the number of balls effectively specifies clearly the conditions under which the probability is to be determined.

But let us go further and turn our attention to one of the peculiarities of the problems treated in the *Essays*, the double lot: the first lot decides the urn, the second the ball. In fact this double lot reduces to a simple one; it does not destroy the independence of the trials if the condition is introduced that the ball must be drawn from a single urn several times or the composition of several urns decided by a single lot.

Such a condition occurs in the *Essays* in all schemes where the dispersion is nonnormal or is normal but cannot be explained by constant probability.

Consequently, in these schemes, because of the double lot and the condition indicated, there is no place for independent trials with different probabilities with which we were concerned in the law of large numbers introduced above.

Nor in particular is there any place in these schemes for Lexis's case where the probabilities are given and not determined by lot. However, we meet this case twice in the *Essays*, on page 349 and on page 357; of course, this is explained by an incorrect use of the terms of the calculus of probability.

In connection with page 357, it is necessary to say frankly that this is not Lexis's scheme. And on page 349 the question is incorrectly formulated: he speaks of a lot and at the same time gives its result; if one eliminates the instruction by lot which leads only to a muddle-headed notion, then we shall indeed have Lexis's case but it will not be a double lot.

Next, in connection with Chuprov's book, I should discuss the question, interesting from a scientific point of view, of the possibility or impossibility of supernormal stability, which has apparently been too hastily disposed of by statisticians[5].

The following two statements in the *Essays* are related to this question. On page 284 after the analysis of a certain case Chuprov writes: "In this way we are convinced that increased stability of a sequence of frequencies depends on the absence of 'independence' among the cases united in a single series, more precisely, the operation of conditions under which a deviation of one or several cases in one direction from the mean tends to induce a deviation of others in the opposite direction." And on page 368 we find: "Supernormal dispersion can only arise when the individual phenomena are not 'independent' of each other."

The common thought in the two cases lies in the assertion that supernormal stability is impossible if the condition of independence is retained. The second time this is asserted without proof and the first time with a reference to an example where independence is destroyed. But this example cannot convince us of anything concerning cases where independence is retained. It shows only that when independence is violated the case of supernormal stability can arise. However, it is far from this to the conviction that supernormal stability is conditional on the absence of independence or testifies to it.

Such a conclusion could be reached only after a complete study of dispersion under the condition of independence. But clearly no one has done this up to this time, because otherwise the result would be different.

There is of course the formula of Bortkiewicz and Lexis, but this is a partial formula and on the basis of it we do not have the right to draw general conclusions. It refers only to those cases where the probability does not change within a series; if the probability changes not only from series to series but also within a series then the formula of Bortkiewicz and Lexis falls and conclusions based on it also fall.

Here I shall not study the general formulas which are in fact very simple, but I shall consider only one sequence of independent trials in which Lexis's case of supernormal dispersion is inseparably connected to the case of subnormal dispersion. The second case differs from a well known example only in that it departs from the scheme of a double drawing indicated above.

The indicated combination of two opposite cases is reached, for example in the series characterized by the following probabilities: 100 times 0.6, 100 times 0.4, 100 times 0.6, 100 times 0.4 ... If we combine the observations in series of 100 trials we obtain Lexis's case of supernormal dispersion but if we combine them in series of 200 the dispersion will be subnormal.

It remains to find the law of large numbers in the *Essays*. The general proposition introduced above does not occur but we find on page 276, under the name of Poisson's theorem, the coupling of Bernoulli's simple case with the determination of the probability of events by means of a double drawing. A double lot looked at in isolation can of course be useful in illustrating the transition from some data to others, but to couple it specifically with Bernoulli's case does not help us in looking at the particular theorem.

This is clarified further, as is clear from the references at the bottom of page 275 and even from the words of the author himself:

"This thesis, introduced by Poisson with the aid of mathematical calculations is in fact so obvious without mathematical proofs that some mathematicians even consider Poisson's complicated derivation to be an exercise in higher mathematics without real content, in its way similar to killing sparrows with cannonballs. However, it is impossible to agree with such a judgment; the great theoretical importance of Poisson's generalization completely justifies the expenditure of labor on its comparatively cumbersome mathematical basis, to say nothing of the fact that the analytical methods introduced by Poisson permit us to obtain further valuable results that are far from obvious."

I have cited these words of the author principally in order to remark on an incorrect understanding of mathematical computations: they cannot prove the obvious and they can only destroy the obviousness.

In the given case the reference to mathematical analysis is especially inappropriate since under the general name of theorem in the *Essays* on page 276, conclusions of analysis are confused with their practical applications to which the name "theorem" is inapplicable; in this way one constructs only mathematical hypnotism for the ignorant.

In conclusion I use this occasion to remark that, in my opinion, some of Bortkiewicz's theoretical studies on dispersion deserve greater attention.

---

[1]First published in *The Journal of the Ministry of Public Education*, New Series, part 31, St. Petersburg, 1911, pp. 369–374 (Editor's note).

[2]The *Essays* appeared in a second edition (Editor's note).

[3]Chuprov, A. A., "The law of large numbers and the stochasticostatistical point of view in contemporary science", in the book *Questions of Statistics*, Gosstatizdat, 1960, pp. 141–161 (Editor's note).

[4]The possibility of extending it to many cases of connected trials was established in my article "The extension of the law of large numbers to dependent variables", (*Proceedings* (Izvestiya) *of the Physico-mathematical Society* at Kazan University, 1907).

[5]Czuber, *Wahrscheinlichkeitsrechnung und ihre Anwendung auf Fehlerausgleichung Statistik und Lebensversicherung*, Leipzig, 1903; Bortkiewicz, "Problems and concepts of scientific statistics", *Journal of the Ministry of Public Education*, New Series, 1910, February, St. Petersburg, pp. 364–365.

Appendix 2

A Review of A. A. Markov's Book, *The Calculus of Probability*
(fourth, posthumous edition, Moscow, 1924)[1]

by
A. A. Chuprov

A. A. Markov's researches on the theory of probability have outstanding interest not only for mathematicians but also for statisticians trying to deepen their understanding of the general foundations of the methods they use in scientific investigation. The new edition of *The Calculus of Probability* should receive a particularly warm response from statisticians since the additions in this edition tend to meet the questions of statisticians directly. The fourth edition of Markov's book *The Calculus of Probability* can, even more appropriately than the preceding edition, be considered a handbook of the theory of probability for statisticians. Neither in the Russian nor in the foreign literature is there a book that could lead so well into the depths of the contemporary formulation of the problems of the theory of probability that are basic for the statistician, or one that is so well acquainted with the most highly perfected methods for their solution.

The characteristic traits of Markov's handbook are sufficiently well known: the sustained rigor of proof, the transparent clarity of the exposition, the carefully weighed reliability in the exposition of the mathematical arguments, the carefully regulated pace of the scientific and creative thought even for the poorly prepared reader, giving many pages the character of a monographic investigation and laying on the whole book the imprint of the author's individuality. So strong a manifestation of the author's personality is, it is true, attended also by some disadvantages for the reader: the choice of material appears a bit one-sided, little attention is paid to the literature on the questions considered, and the flow of scientific thought that is not in the stream of Markov's own work remains outside of the reader's field of vision. But all this is amply expiated by the immediate sensation of that creative development breathed by the pages of the book.

For the statistician the most important chapter is the third, entitled "The law of large numbers" and, in it, paragraph 19 containing a systematic exposition of the principal results of Markov's work on the question of the possibility of extending the law of large numbers to the case of dependent trials. The paramount importance of Markov's generalizations and the methodical completeness of the methods of solution he found for the problems testify to his labors devoted to the problem of the law of large numbers, a concept that is the cornerstone for the probable renewal of the general theory of statistics occurring today. Scattered in separate articles in a series of special journals, Markov's investigations were not easily accessible to a broader circle of readers. Their condensed presentation in a handbook intended for a large market (an edition of 5000 copies) makes them, one can hope, generally accessible not only to specialists in mathematics but also to those statisticians, increasing in number from year to year, who are not terrified at the sight of algebraic formulas. The paragraphs devoted to the derivation of the limit theorems of probability have a more special character but no lesser scientific value. However, in this part of the book one feels especially keenly the author's lack of attention to the elaboration of this question and related questions by researchers who proceed by different paths.

The seventh chapter ("The method of least squares") has great interest for the statistician, both because of Markov's general approach to the construction of the theory of the method of least squares and because of a series of special digressions included in this chapter. For example, in paragraph 45 we find an exposition of the theory of Lexis's dispersion coefficient including Markov's proof that, in the case of normal stability the mathematical expectation of the dispersion coefficient is exactly equal to one. Markov's theory of correlation is studied in the same chapter.

Unfortunately, his form of presentation of the theory of correlation cannot completely satisfy the statistician: the choice of questions on which attention is concentrated is fortuitous, their treatment within the bounds of the chapter on the method of least squares is incomplete, the connection made between the theory of correlation and the theory of probability is inadequate, as in the work of English statisticians, because it is placed in such close contact with the method of least squares. It would correspond better to the spirit of *The Calculus of Probability* if the theory of correlation were not joined to the method of least squares but rather to the construction of Chapter 3, the extension of which, from the case of a single random variable to that of several dependent random variables, constitutes one of the central problems of the mathematical theory of correlation.

In order to work through Markov's book *The Calculus of Probability* in its entirety and to derive from it all that it is capable of giving, a solid mathematical preparation is necessary. However, many pages of the book, among them paragraph 19 which is of greatest interest to the statistician, are accessible also to those who have only a superficial knowledge of an algebra course. I would emphatically advise statisticians whose mathematical equipment does not go beyond the bounds of elementary algebra not to trouble themselves over the enormous difficulty of the greater part of the book but rather to read thoroughly and think through the following chapters, paragraphs, and pages: Chapter 1 pp. 26–44 of Chapter 2; in Chapter 3 paragraphs 14, 15, 16, pp. 98–102 from paragraph 17, paragraph 19; in Chapter 4 paragraph 24; in Chapter 6 paragraph 39 and pages 298–303 of paragraph 40; and if it is not too burdensome, also from Chapter 7, paragraph 43, pages 327–340 from paragraph 44 and pages 345–349 from paragraph 45.

To master this material will require appreciable effort from people who are little accustomed to mathematical computation but the expenditure of time and effort will be repaid a hundredfold. Most instructive will be, as remarked above, Chapter 3 and, as its culmination point, paragraph 19 (supplemented by paragraph 24 of Chapter 4, especially p. 174).

The results of Markov's work expounded here on the extension of the law of large numbers to the case of dependent trials has not yet entered a single Russian or foreign handbook. Even Professor L. K. Lakhtin's[2] *Course of Probability Theory*, which came out almost simultaneously with Markov's (Moscow, 1924) and was intended especially for statisticians, does not contain them.

A portrait of Markov and a "Biographical sketch" by A. S. Besicovich have been added to the fourth edition. The biographical sketch acquaints us with the content, character, and significance of Markov's work but gives no picture of Markov's personality, the report of biographical details taking the form of a service list. The main problem is the incompleteness of the bibliographic information: there is no list of Markov's works appended to the sketch; even the works cited in the biographical sketch are introduced by stating the year and, for the most part, the title but without any indication of where they were published.

It is impossible not to comment on still another inadequacy of the biographical sketch: the information on the relation between Markov's scientific work and the work of other investigators on the same problems hardly goes beyond those narrow bounds to which it is restricted in the works of Markov himself. As a consequence, in places the reader gets an impression that it is simply false. Thus, for example, concerning the theory of dependent trials, the author of the sketch asserts that "it has been scientifically treated only in the works of A. A." Without quarrel, the theory of dependent trials owes its development to A. A. Markov more than to anyone else. One can even say that it was founded by Markov. But it is far from true that it has been worked out only in Markov's works. Not to mention the work of those who are associated with Markov, there are works in this field that are independent of Markov, such as the investigations of Bohlmann and of the Japanese mathematician M. Watanabe;

it is true that in many respects the latter only repeat Markov (and sometimes also Bohlmann) but at the same time they move the development of the theory of dependent trials forward substantially. And in addition, in recent years, physicists going their own way have given much of their attention to the problem of dependent events (Smoluchowski[3] and his concept of probabilistic after-effect).

The picture of Markov as a scientist would have been much sharper if the scientific background against which he must be considered had been described in more detail.

---

[1]First published in *Russian Economic Review* (sbornik), Prague, 1925, February, pp. 171–173.

[2]Lakhtin, L. K.,—Soviet mathematician, author of works in the field of mathematical statistics and probability theory (Editor's note).

[3]Smoluchowski, Marian (1872–1917)—Polish physicist. Parallel with Einstein, he founded the theory of Brownian motion (Editor's note).

Appendix 3

The Bicentennial of the Law of Large Numbers[1]

by
A. A. Markov

Highly esteemed members of this meeting!

From the mathematical point of view, under the name of the law of large numbers in the wide sense we can include the totality of limit theorems of the calculus of probability that fall into two groups of theorems or, if you wish, into two theorems, but with varying conditions. The first group, which consists of the law of large numbers in the narrow sense, and to which my speech will for the most part be devoted, shows us probabilities arbitrarily close to certainty, expressed by the number 1. In the second group the limit of the probability is the well-known integral of Laplace. I shall also say a few words about theorems in this second group, having in view their relation to Jacob Bernoulli's theorem which, in its simplest form, gave rise to the whole collection of theorems united under the name "law of large numbers" and among which, one can say, its extensions occupy the principal place.

J. Bernoulli's theorem was published in 1713 in *Ars Conjectandi* but of course he had discovered it much earlier. Jacob Bernoulli died in August, 1705 and his work *Ars Conjectandi* was published eight years after his death by his nephew Nicholas Bernoulli. However, already in letters to Leibnitz dated 3 October, 1703 and 20 April, 1704 he said of his theorem that twelve years earlier he had shown the proof to his brother John and that the latter had found it to be correct: "Dixi autem in istis me posse demonstrare : viditque demonstrationem jam ante duodecennium Frater et approbavit." Finally, in *Ars Conjectandi* he moved back to twenty years earlier, the time when, if he had not proved, he had at least started to prove his theorem. Here are his words: "Hoc igitur est illud Problema, quod evulgandum hoc loco proposui postquam jam per vicennium pressi."

We can not establish the year in which Bernoulli arrived at his theorem and we do not connect today's celebration to that year but rather to the year his theorem was published, that is, to the year in which *Ars Conjectandi* appeared. That year, 1713 is shown on the book itself and is corroborated by two letters of Nicholas Bernoulli to Monmort that appear in the second edition of Monmort's *Essay d'analyse sur les jeux de hazard*, which appeared at the end of the same year, 1713. In a letter of 23 January 1713 he mentions that *Ars Conjectandi* was being published and that it contained a particular theorem which he justifiably compared with his own calculations, but the latter did not have the force of a real proof. In a letter of September 9 of the same year, Nicholas Bernoulli stated that this work had just appeared. (The letters of John and Nicholas Bernoulli to Leibnitz, containing information of the publication of *Ars Conjectandi*, are marked with the same date). He added that Monmort would not find anything new there.

But this opinion of N. Bernoulli does not prove that Monmort already knew Bernoulli's theorem completely. It is explained by the fact that N. Bernoulli did not attach much importance to the result or considered information given Monmort from memory in a previous letter, where the theorem had been outlined, to be sufficient for him.

Jacob Bernoulli's theorem can be formulated in the following way: "If an unbounded sequence of trials is carried out and in each of these trials a certain event has one and the same probability, then for a sufficiently large number of them one can assert with probability arbitrarily close to certainty that the ratio of the number of occurrences of the event to the number of trials differs from the probability of the event by less than a given number, no matter how small it may be." This theorem and its proof are found at the end of the fourth part of *Ars Conjectandi*, a Russian translation of which has now been published by the Academy.

From J. Bernoulli's comments preceding the proof of the theorem it is evident that he considered it very important, regarding it as the basis for determination of probabilities from observation—a posteriori.

He illustrates his theorem by the following example. White and black balls are mixed in an urn. The ratio of the number of white balls to the total number of balls in the urn is 2/5 and for black balls the corresponding ratio is 3/5 so that the probability of drawing a white ball from the urn is 2/5 and black 3/5. Now let us suppose that the ratios, in other words these probabilities, are unknown, and we can only proceed with an unbounded number of trials each consisting of the drawing of one ball. We write down the colors of the balls drawn in order to keep a record and also the count of the number of white and black balls drawn, and the balls themselves are constantly returned to the urn in order to keep unchanged the number of white as well as the number of black balls in the urn. The question is whether, from this record establishing the ratio of the number of white balls drawn to the total number of balls drawn, we can expect to get arbitrarily close to the unknown probability of a white ball.

Bernoulli's Theorem gives a positive answer to this question. In particular, according to Bernoulli's calculations, one who knows that the white balls constitute 2/5 of all balls in the urn can, with probability differing from certainty by less than 1/1000, assert that in 25,550 trials, that is, after that many balls have been drawn and their colors recorded the ratio of the number of white balls drawn to the total number of balls drawn will lie between 19/50 and 21/50, in other words, will differ from 2/5 by less than 1/50. And if the number of trials is increased, the probability of the assertion can approach arbitrarily close to certainty, that is, instead of 1/1000 can become 1/10,000 or 1/100,000 and at the same time the allowed deviation from 2/5 can be taken to be arbitrarily small, that is, 1/50 can be changed to an arbitrary smaller number.

Jacob Bernoulli's proof is elementary and rigorous but is connected with a limiting condition on the number of observations.

Closely connected with Jacob Bernoulli's theorem is the question of the computation of the probability that the difference between the ratio of the number of occurrences of the event to the total number of trials and the probability of the event does not fall outside of given bounds. It is not difficult to give an exact formula for this probability. Computation by the exact formula reduces to simple arithmetic operations. However, in the case of a large number of trials these computations become quite burdensome and one can say, even impracticable. Thus it is then appropriate to turn to approximate formulas that simplify the computation and shorten it. An attempt to simplify the use of the exact formula, changing it into an approximate one is already found in Nicholas Bernoulli's letter to Monmort dated 23 January, 1713. It is concerned with the interesting question of the stability of the distribution of newborn infants by sex. But it is impossible to attach much importance to it except to observe that it drew the attention of De Moivre[2] (with whose name the well-known trigonometric formula is associated) to the question of the computation of the probability.

With the assistance of Stirling[3], De Moivre succeeded in obtaining, for the probabilities concerning us (at least in the simplest case when the probability of an event equals 1/2) an approximate expression in the form of that integral which we now call the integral of Laplace. De Moivre's derivation can be found in his work *Miscellanea Analytica* of 1730.

In particular, Laplace and Poisson, whose works date from the end of the eighteenth and the first half of the nineteenth centuries, elaborated methods of approximate computation of probabilities in general. I recall that the first edition of Laplace's classical work *Théorie analytique des probabilités* appeared in 1812 and the second in 1814 while Poisson's *Recherches sur la probabilité des jugements en matière criminelle et en matière civile* appeared in 1837.

The question of approximate computation of probabilities is not the subject of my speech. I mentioned it only because the approximate computation of the probability with an appropriate bound for the error can lead to a limit theorem. Thus, for Bernoulli's theorem it yields Laplace's proof with the aid of an inference from the simplest case of the second limit theorem. This simplest case is related to the same difference as in Bernoulli's theorem and we are also concerned with the probability that this difference lies between specified bounds. However, in Bernoulli's case we look at fixed bounds while in the second limit theorem these bounds are proportional to the square root of the reciprocal of the number of trials which, as before, is assumed to increase without bound. The theorem asserts that for a sufficiently large number of trials the probability of not exceeding these bounds becomes arbitrarily close to Laplace's integral.

Combining the second theorem with Bernoulli's theorem can lead, so to speak, to the second stage of Bernoulli's theorem. I shall not stop to formulate it but I shall give some indications, on the basis of which it is easy to do so. Replace each separate trial by a whole collection of trials, the number of which can grow without bound. Then look at an unbounded sequence of such collections. Finally, instead of the original event, look at a new one consisting of the event that the results of the defined collection do not exceed the indicated bounds. Then Bernoulli's theorem reduces to its second stage in that the probability of the original event is replaced by a limiting value of the probability of the new event, that is, by Laplace's integral.

Poisson used the approximate computation for another purpose: for the generalization of Bernoulli's theorem. The name "law of large numbers" is also due to him. I shall not speak about the thought Poisson himself gave to this name and what importance his investigations have for statistics; I shall only discuss the theorem which at the present time mathematicians call Poisson's theorem and the law of large numbers. It differs from Bernoulli's theorem in that the probability of the event is not assumed to be the same for all trials but can have its own particular value for each trial. With such a change of the basic condition, in order to go from Bernoulli's theorem to Poisson's theorem we need only replace the common constant probability of the event in the concluding words of the theorem by the arithmetic mean of the probabilities. Poisson did not prove his theorem because he confined himself to an approximate computation, not bounding its error in an appropriate way.

The first proof of Poisson's theorem was given in 1846 by the unforgettable P. L. Chebyshev in a short but remarkable note "Démonstration, élémentaire d'une proposition générale de la théorie des probabilités" in the 33rd volume of Crelle's Journal under the heading "Extrait d'une mémoire russe sur l'analyse élémentaire de la théorie des probabilités." However, we do not find such a proof in any other work of Chebyshev. One can only conjecture that it is taken from unpublished memoirs or from Chebyshev's master's dissertation "An attempt at elementary analysis of the theory of probability" but that it was not left in there.

I mention in passing that in the same year 1846 the beautiful work of another deceased academician, Buniakovskiĭ's *Foundations of the Mathematical Theory of Probability* appeared.

Twenty years after the first, Chebyshev gave a second proof of Poisson's theorem. It appeared in the Russian language in *Matematicheskiĭ Sbornik* in 1866 under the title "On mean values" and in the French language in Liouville's Journal for 1867. This second proof, based on the consideration of the mathematical expectation of a square is notable for its startling simplicity and yields a more general theorem than Poisson's theorem since here it is no longer a question of the number of occurrences of an event but rather of a sum of different variables. However, it is necessary to remark that his starting point had already been indicated in 1853 by the French mathematician Bienaymé in the memoir "Considérations a l'appui de la découverte de Laplace sur la loi des probabilités dans la méthode des moindres carrés" which was written in connection with an argument between Bienaymé and Cauchy on the advantages of the method of least squares. This memoir appeared in due course in Volume 37 of the *Comptes Rendus* and later was republished in Liouville's Journal for 1867, just before Chebyshev's memoir, but without any indication of the substantial connection between them.

Subsequently Chebyshev, in a short note presented in August, 1873 at a conference in Lyon and also published in Liouville's Journal for 1874, noting this connection, called his second proof a consequence of the new method that Bienaymé gave in the memoir I have cited.

This method, the method of moments or mathematical expectation can be characterized in the following way: We look at the mathematical expectations of various functions of a certain random variable and draw conclusions from them about the probability of these or other assertions concerning it. Although Chebyshev ascribed this method to Bienaymé I consider it more correct to call it the Bienaymé-Chebyshev method, or sometimes simply Chebyshev's method since there is only a rudimentary form of it in Bienaymé's memoir and it acquires significance only through Chebyshev's work. First, Chebyshev connected this method with a special kind of problem concerning maxima and minima similar to problems in the calculus of variations but with the replacement of the continuity condition on the underlying function by a positivity condition corresponding to the fact that masses and probabilities cannot be negative. Second, Chebyshev showed that the method of moments can lead not only to the first but also to the second limit theorem.

I also wish to observe that Bienaymé died in 1878 at the age of eighty-two. In the *Comptes-Rendus* of 1878 we find the obituary by Gurneri. In it are cited the words of Lamé who, in 1851, called Bienaymé almost the only representative of the theory of probability in France. His argument with Cauchy is recalled as well as Bienaymé's memoirs pertaining to this dispute which appeared in the *Comptes-Rendus* in 1853 and in Liouville's

Journal for 1867. It is also recalled that Bienaymé knew various languages and that in 1858 he translated a memoir of Chebyshev into French. But there is not a word of his having developed a new method.

I consider the further development of the law of large numbers to belong already to our time. I shall not expand on it. I shall say only that it consists in the broadening of the field of application of limit theorems and, in particular, in their extension to dependent trials and dependent variables where not only is the method of Laplace and his followers successfully applied but also that of Chebyshev.

In concluding this speech, I return to Jacob Bernoulli. His biographers recall that, following the example of Archimedes he requested that on his tombstone the logarithmic spiral be inscribed with the epitaph "Eadem mutata resurgo". This inscription refers, of course, to properties of the curve that he had found. But it also has a second meaning. It also expresses Bernoulli's hope for resurrection and eternal life.

We can say that this hope is being realized. More than two hundred years have passed since Bernoulli's death but he lives and will live in his theorem.

---

[1]A paper given at a meeting of the Academy of Sciences on 1 December 1913 in commemoration of the bicentennial of the law of large numbers. It was first published in the *Journal* (Vestnik) *of Experimental Physics and Elementary Mathematics*, Second series of the 1st semester, No. 3, Odessa, 1914, pp. 59–64 (Editor's note).

[2]De Moivre, Abraham, (1667–1759)—English mathematician, (French, by birth) (Editor's note).

[3]Stirling, James (1692–1770)—Scottish mathematician (Editor's note).

Appendix 4

The Law of Large Numbers in Contemporary Science[1]

by
A. A. Chuprov

Section I

The future historian of human thought, when examining our contemporary epoch of the end of the nineteenth and beginning of the twentieth century, will observe how its characteristic feature, the striving for scientific knowledge, takes on a statistical form. Those fields grow from year to year where human thought, turning away from single occurrences, concentrates on their combined result, on total or mean values. We can say without exaggeration that the development of contemporary science is in the direction of interest in collective phenomena and that soon there will not be a branch of knowledge where, with more or less success, the statistical form of knowledge has not spread its influence.

What is the reason for such a triumph of statistics in our time?

Let us take as a starting point the well-known words of Laplace in which the greatest mathematical genius sketched the naturalist's ideal of knowledge: "A mind which, at a particular moment, knew all forces operating in nature and the relative positions of all the bodies making up the natural world, if in addition it were sufficiently powerful to use these given data in a computation, would envelop in a single formula the movements of the highest celestial bodies and of the smallest atoms; nothing would be uncertain to him, the future as well as the past would be open to his gaze."

This scientific credo is magnificent in its classical clarity and at the same time it is a grandiose program. Who among scientific workers refuses to repeat this profession of faith in the regularity of Nature, a faith to which we are bound by all the successes of our knowledge but who at the same time does not pause to reflect on the exhorbitant sweep of the program? Let us admit that with combined efforts, having divided the work up among ourselves as observers we achieve in time the possibility of following sufficiently accurately the motion of the sun, the moon, and other heavenly bodies, numerous but not innumerable. However, does it not really seem foolhardy to hope that mankind will ever be in such a state that after planets and stars it can in reality trace the innumerable "smallest atoms"? Laplace's formula itself assumes a bound on the practical ideal of knowledge. Side by side with branches where the "astronomical" ideal, as it was called by Du Bois Reymond, really reigns there is a vast area where it has no place, where Nature, as we believe along with Laplace, is subject to law, is determined by law and predetermined, but where our organic brain does not dare to dream even of an approximation to that completeness of knowledge that Laplace's formula demands. Does that mean, however, that the investigator must submissively lay down his hands, humbly admitting his impotence? Or is there another form of knowledge we think other than the ideal?

Let us suppose, taking an example of Bertrand[2], that a rain cloud is moving. The universal mind of Laplace's formula would foresee the place and time of development of each separate raindrop, compute the prediction of all details of its trajectory and the point where it will come into contact with the surface of the earth; in the picture of how nature will look after the rain nothing would remain unclear. But can we really not foresee with sufficient clarity, despite all our inability to follow the movements of the separate drops, that after a strong rain everything around will be wet and that someone who has gone out without an umbrella will return completely soaked? Or another similar example: the height of newly fallen snow in a broad snow field after a heavy snow storm, the folds of snowdrifts against the bushes, buildings, and fences; do we not really perceive this picture as the final mass result, looked at in the large, of the infinitely intricate intermingling of the snowflakes carried along by the whirlwinds, which we also dare not try to follow in detail?

These examples are typical. Under certain conditions the collective result shows regularity, comprehensible to us without the necessity of knowing exactly the paths of all the individual processes that lead up to it. On this basis we can, alongside the astronomical forms of knowledge, confidently place the statistical forms.

These statistical forms of knowledge are acquiring general importance only in our time but they did not originate in our century. Their scientific application originated in the seventeenth century; their cradle is the same Royal Society of London that was the cradle of contemporary natural science. However, for a long time the sphere of application of these methods was restricted to human social phenomena, and even here they were not put forth independently but timidly clung to natural science, trying, if only outwardly, to reproduce its tested methods. It required a good two centuries before they were established in their own individuality and in all their general applicability, and only in the middle of the nineteenth century does their victorious march over the whole field of contemporary science begin.

This is clear. The statistical point of view signifies a repudiation of that concentration on individual events which appears to the naturalist's mind as the ideal of completeness and perfection of knowledge. Human social relations, on which it was first worked out, clearly cannot be studied in the "astronomical" way. The actions of individual people, with their barely perceptible motives, which are also unique events in the life of individual people are too vast in their intricate variety. And at the same time, taken separately they are often of little interest to us; it is important for us to know their total outcome. Let us take for example the extensive field of insurance. There is no argument: the total damage to insured buildings observed in the final count is only the sum of separate cases of fire and is more closely determined by the fate of the individual insured buildings during the period considered. The insurance companies count on this, systematically influencing the choice of their risks and encouraging the application of fire-prevention measures, but beyond these bounds to the correct allotment of insurance it is largely irrelevant who among the insured suffers and who escapes the bitter cup. If only there is the possibility of foreseeing with sufficient accuracy the total outcome, the organizers of an insurance enterprise would not think of worrying that they cannot tell in advance for which of the insured buildings it will be necessary to pay some compensation.

Under such conditions, when there was nothing to lose, but the gain—the possibility of extending the elements of counting and measuring to the field of human affairs—was enormous, statistical forms of knowledge, introduced into science two and a half centuries ago by Graunt, were received with understandable enthusiasm and quickly took firm root in the social sciences. But it was not so in the natural sciences. Apparently the superficiality of statistical knowledge repelled the natural scientist from the first moments. Only reluctantly, resting at every step did he reconcile himself to a new form of scientific thought, so sharply contradictory to his customary habits. The first sciences to yield were those branches of natural science that are similar to social science in that individual occurrences are almost inaccessible to scientific investigation, disciplines having to do with man himself—anthropology and anthropometry—or with the no less capricious behavior of the weather—meteorology. Later the statistical point of view invaded those branches of biology where man is not the center of attention. Before our eyes it is taking possession of physics, it is beginning to force its way into chemistry and, not without success, is encroaching even upon astronomy.

Section II

I shall not go into the details of the history of the victory of the statistical point of view in all of these varied fields of science. I shall dwell only on a few characteristic examples that elucidate the nature of this in different aspects.

Let us begin with *physics*. By leafing through the physical journals for recent years it is easy to convince ourselves of the extent to which the statistical point of view predominates in physics. At every step we encounter the word "statistics" in the most varied combinations: statistical mechanics, statistical theory of gases, statistical theory of heat and electricity, statistical theory of radiation, and so on. We find this enthusiasm for statistics not only in abstract mathematical constructions but also in experimental investigations: the problem of determining the value of a mass we are studying, technical questions of recording observations, of course each in its own way, concern the experimental physicist as well as the head of a statistical bureau. Not so long ago the statistical point of view, hardly going beyond the kinetic theory of heat, seemed to the majority of physicists to be an audacious game of the mind, quite distant from real scientific problems. But now statistics is a basic method of work without which, one can say, you cannot take a step.

What caused this change in frame of mind?

Let us recall the path by which the statistical point of view first entered physics. The idea that heat was not a substance but a special form of motion—that is, a rapid movement of molecules invisible to the eye—led to statistics by the force of logic. "When we are concerned with masses of matter", explained Maxwell when he introduced the word "statistics" into physics, "the impossibility of observing the individual molecules requires us to take what I have called the statistical method of computation and to renounce the dynamic method in which we follow each separate motion in·our computations."

Molecules and their motions are elusive not only to the eye. Their extreme multiplicity makes them elusive also to the mind. Consequently, in order for kinetic theory to become a working hypothesis rather than merely a pretty picture, it was necessary to find a method, directly accessible to observation, by which the mass results of such elusive motions could be taken into account. Starting from simple assumptions on the nature of molecular motion, statistical mechanics explains their mass results: the uniform pressure of a gas on the walls of its container and the increase of pressure with compression or increasing temperature, and many others.

A concept is obtained that is attractive in its aesthetic completeness and breadth. But some of the conclusions to which it leads in its logical development are striking in their paradoxical nature. The basic laws of physics, firmly established by experiment, prove to be, in light of this hypothesis, not necessary but only highly probable. Yes, a gas presses on the walls of its container uniformly—says the conclusion from the kinetic hypothesis, in agreement with experiment. But this is only a most highly probable result of the bombardment of its walls by the innumerable particles making up the mass of the gas. If the collisions of the individual particles on the wall are really very numerous then, according to the law of large numbers, even the slightest deviations from the most probable result are extremely improbable. But still they are not impossible.

A constellation of motion in which the pressure will be higher at some points and lower at others is also conceivable. We do not encounter this in practice only because of the basic principle of the logic of the probable: the extreme rarity of events of small probability. However, it is only necessary to wait long enough for even the most unlikely combinations to occur. However, one may have to wait myriads of millenia since in practice we always deal with extremely large numbers. Consequently, if in the short period of human existence such exceptional occurrences have not been observed, it is inappropriate to consider this a contradiction to the kinetic theory.

Such a reconciliation of theory and experience is, of course, free from internal contradiction. However, it cannot fail to leave a dull feeling of resentment on the human spirt. Add to that the original sin of the theory, the hypothetical nature of the very existence of those elementary particles the large scale consequences of whose movement are the subject matter of statistical mechanics. It is easy for a social scientist to be a statistician: his atoms, which he does not have the power to follow in detail, are people who exist as independent individuals before his eyes; the phenomena that interest him are made up exactly—he sees it with his own eyes—of the overall results of the things that happen to these individuals. For him there is only the question of how to connect the results to the individual components. The situation is different for the physicist. For him, not only is the connection problematical; even the existence of the components is doubtful. Under such conditions one should not be surprised that after the first enthusiasm for the distinctive beauty of the new theory came the time for disappointment and cooling.

The twentieth century brought radical changes in the situation. Those particles that we are concerned with have now lost their speculative hypothetical character. The physicist has learned how to observe them not in their mass consequences but as individuals, has learned to count them as a statistician counts the inhabitants of a country in a census, has learned how to measure their characteristic traits. "Atoms," according to the contemporary view of Planck,[3] "are neither more nor less real than the heavenly bodies or the inanimate objects that surround us."

These experimental successes are changing everything at once. If the smallest particles are not a play of the imagination but just as real as the heavenly bodies, then for the physicist the statistical point of view ceases to be a matter of individual taste; as for the social scientist it becomes a necessity because of the factual impossibility of following the movements of all these myriads of particles. Like it or not, willy-nilly he must become a statistician if he does not want to exclude from his field of consideration the most important phenomena underlying his subject.

In addition, even those paradoxical conclusions of kinetic theory that were at first so perplexing to scientists are beginning to find experimental corroboration. The uniform pressure of a gas on its container, as the mass result of the continuous bombardment by the particles striking it, turns out to be, as we have seen, not inevitable but only highly probable. One can also expect some deviations, with vanishingly small probability when there are many molecules and the collisions are frequent, but with much larger probability when there are not so many molecules. This naturally leads to the thought of seeking out conditions under which the probability of a deviation has an appreciable value. A long time ago it was already conjectured that such are the conditions of small particles immersed in liquid that describe the odd motions noticed by the English botanist Brown[4].

In recent times the successes of experimental art have permitted us to place this conjecture beyond doubt. Perrin's brilliant experiments showed, one can say, before our eyes, that the translational and rotational motions of such small particles and their distribution in the fluid is directly determined, under the combined action of gravity and random collisions, by a similar kind of random deviation from the most probable state of complete mutual compensation of the individual collisions. And a whole series of diferent experiments in different fields of physics has shown no less obviously that such conclusions from the laws of physics, in the words of Smoluchowski, are not the ravings of the downcast imagination of a visionary theoretician. They are rather, real phenomena which we meet constantly in reality and which are difficult for us to perceive only because of the technical difficulty of observing sufficiently small groups of molecules and atoms.

It is clear how the advocates of the statistical point of view must have been inspired by these experimental successes which showed that they stood, in Rutherford's expression, "on the dependable foundation of reality and not on the shifting sands of suppositions." Now, as a result, we also see bold flights upwards and in breadth. The atomic point of view is being spread further and further. The basis of an atomic theory of matter has been established. After it there appeared on the scene an atomic notion of electricity which, one can say, now prevails. An atom of magnetism, the magneton is trying to join the atom of electricity, the electron. Planck has come to the conviction that the process of radiation of energy cannot be looked upon as continuous. Einstein does not stop even at this. Knowing no restraint, his creative fantasy develops the notion of atomic structure of energy itself, of light in an absolute vacuum and the building is crowned, finally, by the notion of atomic structure of time. "Formerly they taught that nature does not make jumps," explained Lodge in his lecture before the British Association in the fall of 1913, "but in the con-temporary view it comes out that nature only makes jumps."

Of course it is not for me to decide the extent to which all these newly constructed scientific concepts of the atom will bear criticism and will, under further experimentation, display that same degree of reality as the molecules and atoms of matter. Even Lorentz and Planck, the founders of the atomic theory of electricity and the radiation of energy can only with difficulty make their peace with the notion of atoms of light. Some people today try to split the electron itself into smaller parts. Here everything is still in a process of fermentation; only the future will show which opinion will survive. But even at the present time it is already clear that the representation of the phenomena studied by physics as the statistical totals of mass phenomena, understandable only in the average statistical result because of the imperceptibility of the separate processes, stands before our eyes as the cornerstone of the physical scientific world view. But at the same time the significance of the law of large numbers as the logical basis of the statistical point of view is growing. It is becoming even more obvious that the statistician-physicist, in contrast to the statistician in the social sciences, firmly stands on Bernoulli's theorem as a base and invariably appeals to it both for the computation of the most probable mass results and for the determination of the possible deviations from this most probable result. In recent times an attempt has even been made (by M. Laue) to apply the statistical point of view beyond the limits of those cases where the condition of mutual independence of the separate events is satisfied, an extremely interesting direction and, if successful, a very promising one but one that encounters enormous difficulties of principle and technique.

Summarizing, we can emphasize the close connection that exists between the statistical point of view and atomism as a basic conclusion from this metamorphosis of physics into statistics. Everywhere that we come to the conviction that the phenomena in which we are directly interested must be looked upon as consequences of an enormous number of hardly perceptible separate processes, the statistical point of view inevitably underlies our manner of thought and work. Thus it was with physics. In part we see the same thing in chemistry which, in its fundamental aspects, is beginning to take on a statistical coloration in connection with the recently developed picture of the atom as a complicated system of electrons, with the physical and chemical properties of a body depending strongly on the number of electrons and their arrangement inside the atom. The nuclear charge is the basic constant and the atomic weight some function of it, no doubt symptomatic and highly interesting but devoid of fundamental importance.

But astronomy serves as the most instructive illustration. The heavenly bodies are not elusive molecules; and it was in their study that the astronomical ideal of knowledge which we earlier contrasted with the statistical approach was established. However, in all those cases where astronomers have occasion to be concerned with an enormous number of such bodies, even they begin to change the astronomical ideal of following each body separately and seek understanding of the observed picture as a whole on the basis of the same ideas and methods that are fundamental to the statistical mechanics of a gas: the latest studies of the starry world, Kapteyn's theory with its grandiose sweep, is an especially sharp example of this evolution of methodological views in astronomy.

A second essentially quite similar path by which the statistical point of view is penetrating contemporary science leads through the notion of *society*. Everywhere that our thought encounters an unorganized but systematic balance of systems that interact in some way but at the same time pursue their own individual goals in life, scientific reflection about the conditions under which such equilibrium is established, in the absence of outside influence directed toward supporting this equilibrium, leads to statistics and to the law of large numbers. How, in the present economic system, based on a broad division of labor operating in an uncontrolled market, is that close balance between demand and supply, allowing the needs of the separate economic units to be met more or less smoothly, achieved? How is the balance between sexes, which is so essential to society, maintained? How is the proportion of men and women among deaths and births maintained at a nearly constant level from year to year? To such questions there can be no other answer than the law of large numbers.

But such a kind of question inevitably arises to the thoughtful observer, not only in relation to human society. We also find similar relationships in organic nature where man is on the side, where not people but rather various specimens of flora and fauna interact. A change of characters. Is it long ago that social sciences played enthusiastically with the notion of organism? But now botany and zoology work with notions of society and association, and at the same time the field of science of organic nature is being penetrated by the same methods of analysis of empirical data that are used by statisticians in the social sciences. For example, mortality tables are computed for horses, the dying out of trees in the forest under different conditions of growth is studied. And all these life expectancies of horses, trees and even molecules or atoms do not in any way differ in their logical character from the life expectancies of men and women, new-born babies and eighty-year olds that are more familiar to us.

It is true that there is an essential difference between mortality of organisms on the one hand and of molecules and atoms on the other: the latter does not depend on age; atoms, in Perrin's expression, do not grow old while for man or horse the mortality rate increases appreciably with age. There is also observed, in its turn, a radical difference between molecules and atoms: the mortality rate of molecules, their speed of decay varies greatly depending on the conditions under which we observe them, for example on temperature; the mortality of atoms on the other hand does not depend on the conditions of their enviromment, their decay proceeds at the same rate at temperatures close to absolute zero and at the highest temperatures. It is only because of this that the life expectancy of an atom could acquire the importance given to it as a basic characteristic of matter in studies of radioactivity. But, however deep these essential differences in the phenomenon of mortality may be, they do not destroy the formal unity in its scientific treatment.

On a somewhat different basis the law of large numbers is penetrating *meteorology*.

What do we mean by the climatological conditions of a place? The temperature is measured in the morning, at noon, and at night on the first of January; the measurement is repeated on the second of January, on the third of January, and so on, day after day, year after year. Then an average of these irregularly varying numbers is taken. Analogous measurements are made on the air pressure, humidity, and so on. Finally a system of mean values is obtained. Together they characterize not the state of the weather at a particular time of a particular day, but rather that which we call the climate of the place.

On what basis do we ascribe to this number such a meaning? Why do we consider it free from the stamp of time? The answer is not difficult. The comparison of such mean values taken over sufficiently long consecutive periods of time shows that they differ little, that in them we really have something inherent to the given place, as such, independent of the time to which the observations refer. If you wish to account for why the statistical reduction of data through whimsically varying curves of separate measurements shows clear contours of such typical means, you inevitably come to the theory of probability, and the law of large numbers appears here too as the logical basis of your method of reasoning.

Now let us suppose that, having received all elements necessary for the characterization of the climate of a series of locations by means of statistical analysis of meteorological observations, you begin to compare the climates of these places which are subject to different influences. You catch a clear dependence of the climate on the distance from the seacost, on the altitude of the place and so on—continental, maritime and mountain climates—a connection which it is impossible to detect without passing through the crucible of statistical analysis of the separate measurements. Here we enter on a new path, the very broad and well trodden path by which statistics has taken root in modern science. Nearly everywhere, in nearly all fields of knowledge we meet such connections between phenomena which cannot be revealed by consideration of individual observations. Willy nilly one must either have recourse to statistical methods of work or leave this whole circle of phenomena outside one's field of vision. But how can we close our eyes to these phenomena when the center of interest often lies exactly in them?

I shall not introduce examples from the field of social science where this role of statistics is clear to everyone. Let us take natural science: Is there a problem in biology more pressing than the question of the laws of heredity? But how do you approach it outside of the statistical form of work? Galton and Johannsen, Mendelists and their opponents, all agree in that they seek solutions on the basis of statistics. We see the same picture in various fields of applied natural science, I mention only agronomy and its field trials. And even astronomy—the field with the greatest persistence in resisting the invasion of statistics—has bowed its head in recent times and begins to have recourse to statistical methods of analysis of observations with the aim of catching the connections between the phenomena it studies.

There is, finally, one more path by which the statistical point of view steals past in current scientific work. The regularity of mass phenomena is reconcilable with the assumption of incomplete determinism of occurrences individually. Consequently, to one who has a tendency toward *indeterminism* in these or other branches of knowledge, a path is now opening that reconciles the spiritual influence with the possibility of scientific work, a concept developed with the greatest consistency in the work of Renouvier, the most outstanding French metaphysician of the nineteenth century.

Section III

We see that the motives for which contemporary science changes from the astronomical ideal of knowledge to the statistical are extremely diverse. Let us try to bring them together in some system.

Let us first consider the case where the aggregates in which the different phenomena are united are not of independent interest but only perform an exploratory service so to speak. When a doctor, wanting to determine the value of a new medical procedure, compares the average mortality for the series of cases in which it has been applied with that which ordinarily occurred earlier; when a biologist, trying to establish the laws of heredity, compares the mean height of children of tall and short parents, the aggregates generated by these phenomena do not in themselves interest the doctor or biologist, they are only means to an end and once this end is attained they, no longer holding the attention of the investigator, break apart into the individual units generating them. Here the mass phenomenon is entirely the work of our hands; it does not exist by itself independent of our consideration. It does not exist in nature as an independent object of cognition; we construct it arbitrarily and systematically, in accordance with our views and aims.

The basic aim here is *to seize the connections* between the phenomena that interest us and the laws of these connections. Far from always can we reach this aim by the comfortable and uncomplicated paths that are designated in logic under the name of the method of induction. The hereditary relation between parents and their offspring exists just as surely as the relation between barometric pressure and height above sea level. However, try to compare the individual measurements and you will find such a motley picture that you can only throw up your hands if you are not inclined to thrust your own preconceived interpretations on the facts. Only in the case where all relations in our field of sight are strictly inseparable is the comparison of individual observations appropriate to accomplish our purpose. But, in the picturesque expression of Bertrand, "in the boundless world of phenomena, one encounters family relationships of all degrees." It is quite often appropriate for the investigator to have before him the most complicated, most delicate, at times even fantastic interlacements. Truly, in the final analysis all of them can be reduced to the basic notion of the inseparable connection between cause and effect. But the trouble is that this theoretical possibility does not free one from the practical necessity of seizing them beforehand in all their true complexity. One can seize them only with the help of statistical methods, uniting the separate cases in an aggregate and concentrating attention on the characteristics of such aggregates, on mass results and mean values. In their most general aspects, these statistical methods lead us to proceed from the hypothesis of independence and, on the basis of the law of large numbers, estimate how some summary measures or other of the aggregates we have generated ought to behave under this assumption. If the real picture differs sharply from the expected one, the hypothesis of complete independence is rejected as not corresponding to the facts and the conclusion is that there is dependence. In the contrary case, the question remains open for the present.

In this way, the law of large numbers lies at the foundation of scientific work wherever the investigator has to do with dependence rather than firm connections. Whether it be the confused relations of social life, the strange fluctuations of temperature in the human body or in the atmosphere, the windings of the mysterious umbilical cord which, with the strength of hereditary succession, binds the next generation in all fields of organic nature, everywhere the logic of inference concerning the relation of phenomena, with the greatest variety of methodological ornamentation, is the same in its basic lines, and the logical strength of the conclusion— whether we recognize it or not—rests in the final count on the theorem of Jacob Bernoulli.

No less scientific interest is represented by that case where the aggregates we are dealing with are not constructed by us from elements arbitrarily strung together in a series, but rather are actually given to us as such. An arbitrary one of the innumerable nebulae that we see in the starry sky; the city of Petersburg as a complicated system of mutual relationships among an aggregate of its inhabitants; Forest Park as a collection of trees and shrubs, bound together because of their proximity by mutual influence and general conditions of growth; these exist in nature as mass phenomena independent of whether they attract the attention of an investigator or not. And these aggregates can at times be used for further ends. But even if they do not play an official role they nevertheless do not lose interest as an independent and distinctive object of scientific study. The problems for study can be diverse. The first of these, the statistical description of the different objects of such kind, would seem not to have any special interest for us since such a statistical count has at first glance few points of contact with the law of large numbers: in order to determine the number of inhabitants of Petersburg by a census one does not have to turn to Bernoulli's Theorem.

However, in reality it is appropriate now and then to be guided by the law of large numbers even at this stage of statistical work. The fact is that in practice we sometimes find it necessary to depart from the formulation of the problem that seems most natural—a complete count of all the units of the population concerning us: the complete count is changed into a sample count in which the investigation covers only part of the total number of individuals. Astronomers in counting stars, doctors in counting the number of blood corpuscles[5], investigators of plankton in working over the material caught by their nets have long since stood consciously and deliberately on this ground in their work since the impracticality if not the direct impossibility of a complete count is quite obvious in these cases. However, even where a complete count is not completely impossible statisticians are beginning more and more to resort to sampling studies because of the saving of labor and expense.

However, is a sampling study a legitimate method of statistical work? Is it capable of giving a sufficiently sharp representation of the fundamental study of the population? Only the theory of probability can give a satisfactory answer to these questions by clarifying those conditions on which the degree of correspondence between a sample and the underlying population depends. The law of large numbers turns out, in this way, to be even here the basis on which—consciously or not—the statistician rests in his work.

Even more interesting from our point of view is the *problem of stability,* as the statisticians say, that is, in the language of physicists, the problem of *statistical equilibrium* or statistical statics. According to the definition of Bohlmann, statistical statics explores conditions under which the externally perceptible properties of an aggregate of a large number of mechanical individuals remain unchanged despite the lively motions of the separate units. How does it happen that myriads of particles of a gas, moving irregularly in all directions with colossal speeds gives as an overall result, certified by our eyes and our instruments, a picture of calm with unchanging pressure and temperature? Why is it that the annual inventory of the Berlin pawnshop invariably shows from year to year nearly the same number of pocket watches and at the same time the average value of the watches pawned there fluctuates by a few pfennigs about a constant mean of a little over twenty-one marks although, from year to year, they are different watches having different values?

The theory of probability gives an answer to such questions. On its basis we can quite clearly imagine, in broad outline, what combination of conditions can produce such a statistical equilibrium. Already through this we gain quite a bit. Unfortunately the attempts to make an explanation more concrete usually run into insuperable difficulties. Not only to find the numerical values of the probabilities to which the problem leads but even to define clearly what these probabilities are turns out to be nearly impossible in most cases, and only the simplest problems are solved completely. Why, for example, does the roulette ball fall almost equally often on the red and on the black? Poincaré, and before him Kris, gave a clear answer to this. The length of the path of the ball depends on the strength of the throw; the strength of the throw, in its turn, varies. If the law is given connecting the strength of the throw with the probability that the throw will be of one strength rather than another, then we can compute exactly the probability of stopping in one portion of the circumference of the roulette wheel.

Let us carry out such a computation, assuming some definite law; for almost any form of law we come to the conclusion that the probabilities of stopping on the red and on the black fields are equal to each other and quite exceptional conditions are required in order that, with equal extent of the red and black fields, the frequency of times the little ball ends up in them should not be approximately equal within the limits predicted by the law of large numbers. Thus, it is this situation, that sometimes we arrive at the same estimates for the probabilities in which we are interested under almost any initial assumptions, that at times makes it possible to carry the analysis to a conclusion, as Poincaré was fond of emphasizing. Some of the constructions of the kinetic theory of gases come to that. As for social science, there one must ordinarily be satisfied with a summary indication even of the possibility of statistical equilibrium and only in rare cases does it happen that one can push on a bit deeper.

The theory of probability, in clarifying for us the notion of statistical stability, even somewhat incompletely, renders us still another valuable service: it permits us to a certain extent to account at the beginning for the *regularity of coexistence*. Our basic notion of the regularity of Nature, that notion which poured out from the eloquent words of Laplace that I cited earlier, assumes only the regularity of connection in time. Even being adjacent in space at a given moment in time is depicted as random, given, not reducible to laws of mutual dependence between phenomena. Each phenomenon is inseparably connected with some previous one after which it inevitably follows, and with some subsequent phenomenon which it invariably calls forth. But what takes place together with it at the same time is not determined by any laws: at a given moment and a given place in the universe the neighborhood is one thing, at another time and another place it may be completely different. Such is our fundamental notion of causal determinism of the course of the universe. But at the same time we constantly see regularity in proximity, we encounter firm bonds of coexistence. Whence do they emerge? The theory of probability sheds some light on this.

Let us begin with a simple example where this role of the theory of probability shows through with complete clarity. The pressure of a gas on the walls of the container surrounding it is uniform. How is this regularity of coexistence of equal pressure at different points on the surface of the container explained? The kinetic theory clearly reduces this question to the laws of mechanics with the help of the theory of probability. Why do particles of gold lie side by side forming a gold mine deposited in the earth's crust, rather than being scattered throughout its whole extent? An answer is possible in two ways. One can shrug one's shoulders and say: that is the way it is because that is the way it is; that is fundamental, there is no explanation of the distribution of matter in the terrestrial sphere.

But we can also recall analogous cases where a similar distribution arises before our eyes. In gold-fields when the gold-bearing sand is washed, the grains of sand are swept to one side while the grains of gold sink together. Why? Here it is not because it was that way from the beginning. Before the washing, the gold and the sand were mixed together in the most picturesque disarray. But when a strong jet of water is directed at the mixture, setting it in turbulent motion, the heavier particles of gold are separated out and the light grains of sand are carried far away. Suppose, moreover, that the motion of each individual grain of sand or of gold is as imperceptible to our eyes as the movement of individual molecules of a gas. Nevertheless, we clearly foresee the final result of the washing. More than that: we see that this same final result is obtained for an arbitrary initial distribution of the particles of gold and of sand in the mass that is washed, with uniform or non-uniform distribution, with some sort of bond of proximity of particles of gold in the initial state as well as in the absence of such dependence.

We see the same picture in the separation of cream from milk in a separator. And here we clearly understand that the complicated movement which, for an arbitrary initial distribution, the rotation of the centrifuge gives to the separate fat and non-fat particles making up the milk, leads to a final state in which the light fat particles come together in the center while the heavier particles go out to the periphery. In a similar way we can also explain to ourselves the origin of deposits of gold and other metals in the earth's crust.

Suppose that some time in the distant past the different substances were distributed in an arbitrary way on the globe. In the course of thousands of years processes analogous to those considered above must have produced their regrouping, sorting out similar particles in one place, at least partially.

I shall not insist on a similar analysis from this point of view of the various relations of coexistence observed by us. For our purposes it will be sufficient to indicate superficially the role of the theory of probability in the system of our understanding of the world.

The last group of problems on which I shall dwell lie already at the boundary of scientific knowledge: this is the field of *scientific prediction*. Fortune-telling—any real scientist waves it aside scornfully. In a way this is right. But all our life is based on such fortune-telling about the future; without it we cannot take a step. And in the highest branches of science the problem of predicting what will happen takes a place that is not the last. Solar and lunar eclipses, the rising and setting of heavenly bodies, the rising and falling of the tides—you see that the logic of scientific prediction is no less worthy a subject of attention for the mind trying to give itself an account of the paths to the attainment of truth than those processes by which the regularity of Nature is disclosed or its expediently stylized representations are created.

By what paths then does thought proceed in the solution of the question of what should be expected in the future? Examining the question carefully, we distinguish, without difficulty, two types of logical processes corresponding to the astronomical form of knowledge and the statistical form that we have contrasted above. The logic of the first kind of prediction is based entirely on Laplace's formulation. If you know the general laws of all that happens in the world and the initial positions of the objects, then the future is open to your view if you dispose of adequately perfected mathematical apparatus. According to this scheme or, more precisely, some approximation to it, a prediction is also made, for example, in astronomy.

But is this how we form our expectations in other fields of scientific and vital interest? Do we often know the initial situation with sufficient precision? Do we always dispose of the necessary knowledge of the laws? And the laws themselves, on which we base our work—do they always conform to that picture that lies at the base of Laplace's formulation? Those connections among phenomena with which we come to grips not rarely have a completely different character: they are not unbreakable connections but more or less close connections: one of the related phenomena follows from the other, not always and everywhere, but only with greater or less probability. In order to base predictions on such a foundation, evidently different methods of reasoning are necessary. And in reality if you begin to analyze how our predictions in fact arise, you will convince yourselves without difficulty that they have few points of contact with Laplace's formulation.

The difference between the two modes of scientific reflection about the future is brought out with extraordinary vividness in physics. The famous second law of thermodynamics says, in its classical form, that in a system isolated from outside influences changes can take place only in a fixed direction, namely, that in which a certain numerical characteristic of the system, the entropy, increases; a change under which the entropy of a closed system would decrease does not occur. If two bodies of different temperature are brought into contact, heat begins to flow from the warmer to the cooler body and never flows from the cooler to the warmer. Two gases diffusing mix together and never, after they have been mixed together, do they separate into the component parts without external influence.

Thus did the matter appear before the penetration of statistics into physics. The statistical point of view has changed everything. Thus the direction of processes which earlier seemed the only possible one is, to the advocates of the statistical point of view, only highly probable, not excluding, they assert, the possibility also of the opposite direction. It is not impossible that when a warm and a cold gas are brought into contact the warm gas becomes even warmer and the cold one gets colder; under a certain random constellation of movement of the molecules of both gases—if in the warm gas the slowly moving molecules are for the most part randomly directed to the side where it is adjacent to the cold gas and in the cold gas, the reverse, the faster molecules fly to the side of the warm gas—the warm gas will get warmer and the cold gas colder. If is true that when we look at any appreciable mass of gas such random behavior is extremely improbable. But nevertheless the probability that this will happen is not 0, and if it is already too much to pin one's hopes on the indications of traditional science, then now and then, though extremely rarely under ordinary circumstances, one can be wrong.

If one wants an astronomically exact prediction in the spirit of Laplace's formulation, then it is necessary to dispose of knowledge of exactly the sort that it assumes. It is necessary to know not only the laws of motion but also the initial conditions of all of the smallest particles. Even the connections between these summary characteristics of the mass of particles, the knowledge of which gives us our ordinary measurements, do not have unbreakable and simple characters and thus do not permit us to predict with certainty exactly what will happen in the future if the present is considered to be adequately described by its summary statistical traits.

At first these conclusions of statistical physics seemed to be paradoxical to the point of unacceptability. However, now it is appropriate to recognize them as irreproachably accurate. They do not in essence shake that confidence which we, under ordinary circumstances, place in the second law of thermodynamics since under these conditions the probability of disappointment is small beyond measure. But at the present time it is possible to make observations under such conditions that this probability is not so insignificant and as a result, it is ascertained experimentally that processes can deviate appreciably from the most probable direction. "We see," Smoluchowski characterizes the contemporary state of affairs, "how small bodies come to move about, how they do work, overcoming the force of gravity or of magnetism; we see how inequalities in density and concentration arise automatically."

In this way we encounter a relation between determinism and indeterminism directly opposite to that sketched by the imagination of Renouvier. The individual phenomena are very strictly determined in all details; but the relations between the summary statistical characteristics of successive mass states are not always unique: one and the same initial state in its macroscopic traits is superceded sometimes by one, sometimes by another following state depending on the variety of microscopic constellations that can correspond to it. The concept of a distinctive *statistical indeterminism,* arises naturally and, one can say, inescapably—by the force of logic—on the basis of the whole-hearted recognition of the complete determinism of the motion of the universe with the transition from the astronomical tracing of the individual processes to the statistical study of the mass. This is a concept which, in a somewhat different aspect, relative to the problem of organic evolution, is brilliantly developed by the French sociologist Tarde in his theory of "évolution multiforme" under the strong influence of the most outstanding creator of the contemporary logic of the probable, Augustin Cournot.

"Savior c'est prévoir"[6] runs the slogan of positive science. We are convinced that both in the field of knowledge and in the sphere of prediction contemporary scientific thought is striving in its deepest aspects to take a statistical form. An ever-increasing portion of the phenomena studied by science is moving away to the category of mass phenomena subject to the authority of statistics. The following question can even arise under the form of the contemporary dominance of the statistical point of view: how did science exist earlier without knowing that it had to do with mass phenomena? How could it grow without resorting to the help of those statistical methods of investigation that follow from the logical nature of mass phenomena? "The law of large numbers" dispels this perplexity also. Those mass phenomena with which the experimental physicist, for example, is concerned generally envelop a large collection of similar isolated cases; because of this even the smallest deviations from the most probable state are unimaginably improbable and so rare that one can even disregard them without special detriment. Brownian motion, which tirelessly throws the microscopically small, heavy spherical particles upward before our eyes, not permitting them to settle tranquilly at the bottom of the liquid, can even lead, as Perrin points out, to a brick on a building site raising itself spontaneously to a stone mason at the height of the second floor; but the probability of this is so insignificant that it would require $10^{10^{10}}$ years for the chance of such an event to be at all appreciable. In practice, not only every-day practice but also scientific—in ordinary conditions of scientific work—one can without risk ignore such complications; one can investigate such mass phenomena forgetting for the time being their true character, one can assume that the stone always goes down in air and sinks in water.

Carried consistently to the end, the statistical point of view itself introduces limits. The individual phenomena on the one hand are often mass phenomena, if one can use such an expression; on the other hand, under certain conditions they can be studied reasonably successfully without the work being directed consciously in a statistical channel. Only the intermediate zone steadily requires the application of well thought out statistical methods for a completely successful treatment. And if at the present time statistics is capturing all fields of science with irrepressible force, the basic reason everywhere seems to be the awakening of interest in phenomena of exactly this intermediate area. It is not for our generation to solve the question of the extent to which here, in its turn, apart from the immediate influence of some branches of science on others, historical chance manifests itself and, insofar as they play a role, the general trends of development of scientific thought; excessive nearness conceals a general historical perspective from us. However, we already have the right to state—in reply to the malicious lamentations being heard from various quarters over the bankruptcy of science—that our scientific knowledge is emerging renewed and strengthened from its past crisis with unshakable basic stability.

---

[1]A paper given at a meeting of the Academy of Sciences on 1 December, 1913 in commemoration of the bicentennial of the law of large numbers. It was first published in *Statisticheskiĭ Vestnik*, 1914, No. 1–2, pp. 1–21.

[2]Bertrand, Joseph (1822–1900)—French mathematician (Editor's note).

[3]Planck, Max (1858–1947)—German physicist (Editor's note).

[4]Brown, Robert (1773–1855)—English botanist who discovered Brownian motion (Editor's note).

[5]Apart from its methodological interest, the problem of counting the number of blood corpuscles deserves to be mentioned in the history of theoretical statistics also because, in 1878, in studying it, the well-known physicist Abbé developed the mathematical theory of the law of large numbers in the case of very small probabilities. His attention had been called to this problem by the manufacture, at the Zeiss factory, of a special instrument for counting blood corpuscles. It had been touched on earlier by Poisson just in passing and was later thoroughly worked out by Bortkiewicz in "Das Gesetz der kleinen Zahlen". Abbé's paper completely escaped Bortkiewicz's attention and it was not mentioned by physicists, but Abbé's formulas found application both in the field for which they were developed and in research on plankton.

[6]To know is to foresee.